MAGNETIC CRITICAL SCATTERING

OXFORD SERIES ON NEUTRON SCATTERING IN CONDENSED MATTER

SERIES EDITORS
S. W. LOVESEY
E. W. J. MITCHELL

1. G. Williams: *Polarized Neutrons*
2. E. Balcar and S. W. Lovesey: *Theory of Magnetic Neutron and Photon Scattering*
3. V. Sears: *Neutron Optics: An Introduction to the Theory of Neutron Optical Phenomena and Their Applications*
4. M. F. Collins: *Magnetic Critical Scattering*

MAGNETIC CRITICAL SCATTERING

MALCOLM F. COLLINS

McMaster University

New York   Oxford
OXFORD UNIVERSITY PRESS
1989

Oxford University Press

Oxford New York Toronto
Delhi Bombay Calcutta Madras Karachi
Petaling Jaya Singapore Hong Kong Tokyo
Nairobi Dar es Salaam Cape Town
Melbourne Auckland

and associated companies in
Berlin Ibadan

Copyright © 1989 by Oxford University Press, Inc.

Published by Oxford University Press, Inc.,
200 Madison Avenue, New York, New York 10016

Oxford is a registered trademark of Oxford University Press

All rights reserved. No part of this publication may be reproduced,
stored in a retrieval system, or transmitted, in any form or by any means,
electronic, mechanical, photocopying, recording, or otherwise,
without the prior permission of Oxford University Press.

Library of Congress Cataloging-in-Publication Data
Collins, Malcolm F.
Magnetic critical scattering/Malcolm F. Collins.
p. cm.—(Oxford series on neutron scattering in condensed matter)
Bibliography: p. Includes index.
ISBN 0-19-504600-5
1. Magnetic materials. 2. Critical phenomena (Physics)
3. Critical scattering (Physics) 4. Thermal neutrons—Scattering.
5. Phase transformations (Statistical physics) I. Title.
II. Series.
QC765.C65 1989
538'.3—dc19 88-17438 CIP

1 3 5 7 9 8 6 4 2

Printed in the United States of America
on acid-free paper

PREFACE

Critical phase transitions are a fascinating field of study. Near the critical point there are large fluctuations of physical properties from their average values and the proper description of the state of the matter is a challenging problem for both experimentalists and theoreticians. Although critical phase transitions occur in a variety of systems with different microscopic interactions, the theory has been developed mainly in a language appropriate to magnetism, since magnetic systems have the simplest set of microscopic interactions.

Part I of this book provides an introduction to the theory of critical phenomena set mainly in the context of magnetic critical phase transitions. The modern era of treatment of such transitions started in the 1940s with Onsager's solution of the two-dimensional Ising model. This solution represented something of a mathematical tour de force at the time and it was soon realized that the production of an analogous exact solution in three dimensions was not likely to be forthcoming. Attention turned to approximate methods and in the late 1950s the idea arose of describing critical phase transitions by critical exponents. The 1960s saw the development of the scaling approach to critical phenomena, and this was taken a step further by the introduction of renormalization-group techniques in the early 1970s. The last decade has seen steady advances in many areas but no revolutionary changes of outlook, so now is a good time to review the subject. The research literature on critical phenomena is formidably large and this book only provides an introduction to the area.

In his original discovery of a critical phase transition in carbon dioxide, Andrews noted that there was excessive scattering of light in the vicinity of the transition. This occurs when the fluctuating critical regions have a size of the order of the wavelength of light. In the magnetic case, analogous critical scattering is found for the slow neutron, a particle that interacts through its magnetic moment with magnetic fluctuations. The neutron probes such fluctuations over length scales from one to a few hundred atomic diameters and has become the prime tool for investigating magnetic critical scattering. Part II of this book develops the theory of the use of the neutron as a means to investigate critical phenomena and describes the practice of the use of neutron scattering in this regard.

Part III of this book gives some highlights of what has been learned by the measurement of magnetic critical scattering. The selection of material is entirely my own. I have included examples of simple materials where the magnetic system is one-, two-, and three-dimensional as well as more complex materials showing multicritical points, percolation effects, random fields, and crossover exponents. There is also a review of experimental measurements of critical scattering in transition metals, for which the nature of the paramagnetic state is not clear.

It is hoped that this book will be of interest to researchers in critical phenomena or in neutron scattering. The level of presentation should be appropriate for use in a graduate course. Part I, in fact, comes mainly from a graduate course that I have taught at McMaster University. At the end of the chapters, I have provided a list of suggested works for further study that are particularly readable and will provide more extensive discussion of the topics concerned.

I would like to thank Bruce Gaulin, Karen Hughes, Hong Lin, and Thom Mason for reading the manuscript and making numerous comments. I am also indebted to John Berlinksky, Bob Birgeneau, Peter Böni, Catherine Kallin, and Izabela Sosnowska for suggestions that have improved this work. I should like to thank Cheryl McCallion and Jane Hammingh for performing the arduous task of typing the manuscript.

Hamilton, Ontario M. F. C.
April 1988

ACKNOWLEDGMENTS

I am indebted to the following for permission to reproduce diagrams: J. Als-Nielsen, A. Arrott, R. J. Birgeneau, P. Böni, R. A. Cowley, R. J. deJonge, J. M. Hastings, K. Hirakawa, M. T. Hutchings, H. Kadowaki, G. H. Lander, C. F. Majkrzak, F. Mezei, V. J. Minkiewicz, R. M. Moon, L. P. Regnautlt, E. J. Samuelsen, S. A. Werner, the American Institute of Physics, the American Physical Society, Elsevier Science Publishers, the Institute of Physics (U.K.), the Physical Society of Japan and Spring-Verlag. Appendix A uses data by permission of V. F. Sears and the International Union of Crystallography.

CONTENTS

I THEORY OF CRITICAL PHENOMENA

1. INTRODUCTION TO CRITICAL PHENOMENA, 3
2. GINZBURG–LANDAU THEORY, 8
3. CRITICAL EXPONENTS, 12
   3.1 The basic idea, 12
   3.2 Inequality relationships between critical exponents, 14
4. UNIVERSALITY, STANDARD MODELS, AND SOLVABLE MODELS, 16
   4.1 Universality, 16
   4.2 Standard models, 17
   4.3 Solvable models, 18
5. SCALING, 20
   5.1 The scaling approach, 20
   5.2 Solution of scaling equations; scaling laws, 22
   5.3 Correlation function approach, 25
   5.4 General features of scaling theory, 28
6. THE RENORMALIZATION GROUP, 31
   6.1 The Gaussian model, 31
   6.2 Beyond the Gaussian model, 36
7. CRITICAL DYNAMICS, 42
   7.1 Dynamic scaling, 42
   7.2 Heisenberg ferromagnet, 44
   7.3 Heisenberg antiferromagnet, 45
8. MORE COMPLEX MAGNETIC SYSTEMS, 48
   8.1 Crossover exponents, 48
   8.2 Tricritical points, 51
   8.3 Ginzburg–Landau theory of tricritical points, 52
   8.4 Bicritical points, 54
   8.5 Lifshitz points, 55

9. DILUTION, PERCOLATION, AND RANDOM FIELDS, 58
   9.1 Dilution, 58
   9.2 Percolation, 59
   9.3 Random fields, 60

II THE TECHNIQUE OF THERMAL NEUTRON SCATTERING AND ITS APPLICATION TO INVESTIGATION OF CRITICAL PHENOMENA

10. BASIC PROPERTIES OF THERMAL NEUTRONS, 63
   10.1 Introduction, 63
   10.2 The cross section, 65
   10.3 Nuclear scattering, 67
   10.4 Separation of the nuclear scattering into coherent and incoherent parts, 68
   10.5 Magnetic scattering, 69
   10.6 Spin-only scattering from unpolarized neutrons, 70

11. CORRELATION FUNCTION FORMALISM, 74
   11.1 Nuclear scattering, 74
   11.2 Magnetic scattering, 76
   11.3 Measurement of static correlation functions; the static approximation, 77
   11.4 Elastic scattering, 79

12 BRAGG SCATTERING, 81
   12.1 The scattering geometry, 81
   12.2 The scattering intensity, 83
   12.3 Monochromators and analyzers, 84
   12.4 Effects of beam collimation and of mosaic spread, 85

13. MEASUREMENT OF CRITICAL DYNAMICS, 90
   13.1 The triple-axis spectrometer, 90
   13.2 The neutron spin-echo technique, 90

III MEASUREMENTS OF CRITICAL SCATTERING

14. TWO- AND ONE-DIMENSIONAL SYSTEMS, 94
   14.1 Introduction, 94
   14.2 Two-dimensional Ising systems, 95
   14.3 Two-dimensional $X-Y$ systems, 100

14.4 Almost two-dimensional Heisenberg systems, 105

14.5 One-dimensional systems, 108

15. ISING SYSTEM IN THREE DIMENSIONS, 111

   15.1 Magnetic systems, 111

   15.2 Ising antiferromagnets, 112

   15.3 The Ising dipolar ferromagnet, 116

   15.4 Ordering in alloys, 117

16. OTHER SIMPLE SYSTEMS IN THREE DIMENSIONS, 121

   16.1 Heisenberg ferromagnet, 121

   16.2 Heisenberg antiferromagnet, 128

   16.3 Transitions with SO(3) universality class, 132

17. MULTICRITICAL POINTS, 135

   17.1 Tricritical points, 135

   17.2 Bicritical points, 136

   17.3 Lifshitz points, 137

18. CRITICAL PHASE TRANSITIONS IN MAGNETIC METALS, 140

   18.1 Introduction, 140

   18.2 Iron, cobalt, and nickel, 141

   18.3 Itinerant transition metals: chromium and MnSi, 149

   18.4 Ferromagnetic transition-metal compounds: MnP and Pd_2MnSn, 154

   18.5 Rare-earth metals, 156

   18.6 Uranium and cerium compounds, 156

19. CRITICAL SCATTERING INVESTIGATIONS OF DILUTION, PERCOLATION, AND RANDOM-FIELD EFFECTS, 164

   19.1 Dilution, 164

   19.2 Percolation, 166

   19.3 Random fields, 167

APPENDIX. Thermal Neutron Scattering Lengths and Cross Section for the Stable Elements and for a Few Selected Isotopes, 170

REFERENCES, 173

AUTHOR INDEX, 181

SUBJECT INDEX, 185

MAGNETIC CRITICAL SCATTERING

1

INTRODUCTION TO CRITICAL PHENOMENA

The existence of critical phase transitions was discovered by Andrews (1869) in his classic studies of carbon dioxide. Two phases, liquid and vapor, became identical at the critical point and there was a seemingly continuous transition from the one phase to the other.

Figure 1.1 shows a plot of the pressure as a function of temperature for such a system. The system shows three phases: vapor, liquid, and solid. There are areas where the system can be said to be in one of these three phases, and these areas are separated by boundary lines known as *phase boundaries*. Two phases can coexist at phase boundaries; the system is found to be heterogeneous with regions corresponding to each of the phases. For instance, on the vapor–liquid phase boundary the system will have part of its volume occupied by liquid and part by vapor with a surface separating the liquid and vapor. Usually gravity will have the effect of pulling the liquid phase to the bottom of the volume.

The point where the three phase boundary lines on Figure 1.1 intersect is known as the *triple point*. All three phases can coexist at this particular temperature and pressure. Each phase will occupy its own local volume with surface boundaries separating one phase from another.

If the system is taken across any of the lines in Figure 1.1, by varying the temperature or pressure, there is a discontinuous change of properties and a latent heat is exhibited. A remarkable feature of Figure 1.1, however, is that the phase boundary between liquid and vapor simply dies out at a point known as the *critical point*. At this point the liquid and vapor become indistinguishable. By going round the critical point in Figure 1.1 on the right-hand side it is possible to take a path from the region marked "vapor" to the region marked "liquid" without crossing any phase boundary. Thus, we can go from the vapor to the liquid without experiencing any discontinuous change in properties and without there being any point at which it can be said that the phase changes from vapor to liquid.

These features can be described equivalently in terms of the energy of the system. The appropriate energy function for the variables P and T is the *Gibbs free energy* $G(P, T)$. When two phases coexist at a phase boundary their free energies must be equal. However, there is no reason why the partial derivatives $\partial G/\partial p$ or $\partial G/\partial T$ need be the same at the boundaries and normally they would be different, so that in crossing a phase boundary there would be a discontinuity in the first derivative of G. Such a transition is called a *first-order phase transition* because the

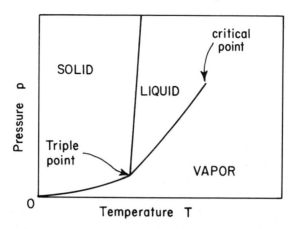

Figure 1.1. Pressure–temperature diagram for a typical gas–liquid–solid system. There are first-order phase transitions between the phases. The point where the three lines representing these phase transitions all meet is known as the *triple point*. The line of first-order phase transitions between gas and liquid states stops at a point known as the critical point. At the critical point (p_c, T_c) the gas and liquid phases become indistinguishable.

first derivative of G is discontinuous. Because of the thermodynamic relationships that the specific heat at constant pressure is given by

$$c_p = -T \frac{\partial^2 G}{\partial T^2}\bigg|_p$$

and by

$$c_p = \frac{\partial H}{\partial T}\bigg|_p$$

where H is the enthalpy, it follows that H is discontinuous at the phase transition and that there is a latent heat involved with the transition.

Phase transitions that pass through a critical point are known as *critical phase transitions*. Historically, such phase transitions were labeled (by Ehrenfest) as second-order phase transitions, in the expectation that they would correspond to a discontinuity in the second derivative of G. However, it became apparent later that this classification was not appropriate, because the behavior of G at a critical point is found to be nonanalytic, so the phase transition was renamed as a critical phase transition.

First-order phase transitions are difficult to understand on a microscopic scale because the physical properties of the two phases are radically different. The entropy difference between the phases is sufficiently large in most cases to preclude statistical fluctuations in each phase that give short-lived microregions of the other phase close to the phase boundary and the phase transition may have to take place by nucleation of the second phase at some special points, such as impurities or boundary walls, followed by growth of the regions of the second phase.

In critical phase transitions, short-lived fluctuating microregions of one phase in the other are always found; they give rise to critical scattering, which is the subject of this book. As the critical point is approached, the size of these regions increases and becomes infinite at the critical point.

We now turn to the magnetic case in which a continuous phase transition occurs from an ordered ferromagnetic state to a paramagmetic state. The critical point is at zero applied magnetic field **H** and at a temperature known as the *Curie temperature*, T_c. The ferromagnetic state is characterized by the magnetization **M**, whose variation with temperature is shown in Figure 1.2. The solid line shows the variation of the magnitude of **M** with temperature in limitingly small field **H**. **M** itself is a continuous function, though the derivatives of **M** diverge at T_c. The phase transition at T_c has no associated latent heat and can be described as a critical phase transition. If one examines the magnetization in high magnetic field, as shown by the broken curve in Figure 1.2, it is apparent that it is possible to pass from the ferromagnetic state to the paramagnetic state without going through any point that can be associated with a phase transition. In other words, we can go round the critical point from one phase to another without experiencing a transition point, just as we can in the liquid–gas case.

We can push the analogy between the two cases further by noting that for the liquid–gas system the energy involves the pressure and volume through a term $-p\,dV$, while for the magnetic case the corresponding term is $\mathbf{H} \cdot d\mathbf{M}$. In the former case the critical point (p_c, V_c) corresponds

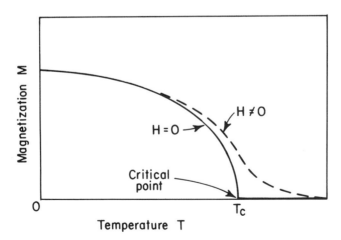

Figure 1.2. Magnetization against temperature for a typical ferromagnet. In zero field there is a critical phase transition at temperature T_c. The presence of an external field H removes the critical phase transition, so that there is a critical point at $H = 0$ and $T = T_c$.

to nonzero pressure and volume, while in the magnetic case it corresponds to zero field and magnetization. An important difference arises because **H** and **M** are vector quantities while p and V are scalars. In Figure 1.2 we have plotted the magnitude of the magnetization **M** against temperature, assuming its direction to be the same as the direction of **H**. Real ferromagnets are not isotropic, because dipolar interactions between atomic magnets are not isotropic and because any interaction between atomic magnets and the local (crystalline) environment will not be isotropic but will have the local symmetry. However, in so-called soft magnets the energy per atom involved in such anisotropies is, maybe, three or four orders of magnitude less than $k_B T_c$ (k_B is Boltzmann's constant), so it is a reasonable approximation to neglect the anisotropy. We introduce the concept of an *ideal magnet* to describe a magnet that is truly isotropic. For such an ideal magnet, the magnetic properties will not depend on the direction of **H**, but only on the magnitude of **H**.

Critical phase transitions with the same general properties are also found for ferroelectricity, superconductivity, superfluidity in liquid helium, mixing of liquids, and ordering in alloys. It is quite amazing that these systems, which have very different microscopic Hamiltonians, all show similar types of phase transitions. The production of a theory that predicts critical phase transitions under all these different conditions represents a formidable challenge! However, as we shall see, this challenge has largely been met.

Let us start by noting three properties that are common to all critical phase transitions.

1. There is a symmetry that is broken at the critical point. This symmetry is represented by a parameter, η known as the *order parameter*, that is a continuous function of the temperature; η is zero for temperatures above T_c and is nonzero for temperatures below T_c. For ferromagnetism the order parameter is the magnetization, while for the liquid–gas transition it is the difference between the densities of the liquid and the gas phases.

   The order parameter may be a scalar or a vector of dimensionality D. It will become apparent later that the dimensionality, or number of degrees of freedom, of the order parameter has a strong influence on the properties of the phase transition.
2. Near the critical point there are fluctuating microregions of both phases involved. The linear extent of these regions (the *correlation length*, ξ) tends to infinity as the critical point is approached from any direction.
3. The response time of the system tends to infinity as the critical point is approached from any direction. This is known as *critical slowing down*.

It is believed that the last two of these conditions are a consequence of the first condition and of Goldstone's theorem (Goldstone et al. 1962), which states that any symmetry-breaking process has associated with it a boson of zero mass. Such a boson needs limitingly small energy to be created and, in the ideal magnet, corresponds to a transverse fluctuation of the magnetization (for $D > 1$); that is, the magnetization vector can be rotated at zero cost in energy.

Suggested Further Reading

Andrews (1869)
Stanley (1971)

2

GINZBERG–LANDAU THEORY

The Ginzberg–Landau theory (also sometimes called the mean-field theory or just the Landau theory) is the simplest treatment of continuous phase transitions and has the merit of being solvable in just about every case. Its roots lie in the Van der Waals theory of the liquid–gas phase transition, in the Weiss molecular-field treatment of magnetic phase transitions, and in the Bragg–Williams theory of ordering in alloys. Landau showed that these early theories were based on analogous assumptions and generalized the ideas to all continuous phase transitions.

The basic assumption of the theory is that the thermodynamic energy functions can be expanded as Taylor series near the critical point. This assumption is in fact not correct, as the functions are not analytic at the critical point. However, the theory does hold for interactions of infinitely long range and for systems with four or more dimensions. Its simplicity and generality make it a good starting point for treatments of continuous phase transitions, and it allows us to introduce some concepts and frameworks that will be needed later.

Let us develop the theory for a phase transition from a ferromagnetic to a paramagnetic state. Then the order parameter is the magnetization and the energy function appropriate for the variables T and M is the Helmholtz free energy $F(T, M)$, or, what is the same thing, $F(T, \eta)$. Near the critical point the order parameter will be small and we expand to get

$$F(T, \eta) = F_0(T) + \alpha_2(T)\eta^2 + \alpha_4(T)\eta^4 + \cdots \qquad (2.1)$$

where $\eta^2 = \boldsymbol{\eta} \cdot \boldsymbol{\eta}$. Because the free energy is the same for a magnetization $\boldsymbol{\eta}$ and for its negative $-\boldsymbol{\eta}$, odd terms will not appear in the expansion.

At a given temperature the system will be in equilibrium when $F(T, \eta)$ has a minimum value. That is, when

$$\left.\frac{\partial F}{\partial \eta}\right|_T = 0 \quad \text{and} \quad \left.\frac{\partial^2 F}{\partial \eta^2}\right|_T > 0$$

This implies

$$\eta\alpha_2(T) + 2\eta^3\alpha_4(T) = 0 \qquad \alpha_2(T) + 6\eta^2\alpha_4(T) > 0$$

Above the critical temperature T_c we requires that $\eta = 0$, so that

$$\alpha_2(T) > 0 \qquad (T > T_c)$$

Below the critical temperature, η is not equal to zero (else there would not be a phase transition), so

$$\alpha_2(T) = -2\eta^2 \alpha_4(T) \quad (T < T_c)$$
$$\eta^2 \alpha_4(T) > 0 \quad (T < T_c)$$

so that

$$\alpha_4(T) > 0 \quad \text{and} \quad \alpha_2(T) < 0 \quad (T < T_c)$$

There is a change of sign of $\alpha_2(T)$ at the critical temperature. If we expand $\alpha_2(T)$ about $T = T_c$ and retain just the lowest nonzero term, we have

$$\alpha_2(T) = (T - T_c)\alpha_0 \tag{2.2}$$

with α_0 positive, and get

$$\eta^2 = \frac{\alpha_0}{2\alpha_4(T)}(T_c - T) \quad (T < T_c) \tag{2.3}$$

and

$$F(T, \eta) = F_0(T) \quad (T > T_c) \tag{2.4}$$
$$F(T, \eta) = F_0(T) - (\alpha_0^2/4\alpha_4(T))(T_c - T)^2 \quad (T < T_c) \tag{2.5}$$

Figure 2.1 shows this result graphically. Above the critical temperature the free energy has a minimum at $\eta = 0$, while below the critical temperature there are minima at $\eta = \pm[\alpha_0(T_c - T)/2\alpha_4(T)]^{1/2}$. For a vector magnetization, we would take the minimum below the critical temperature as corresponding to the magnitude of the vector $\boldsymbol{\eta}$, with the direction arbitrary.

We can use this solution to calculate the specific heat. Of the two magnetic specific heats, c_M and c_H, it is the latter, the specific heat at constant field H, that is most easily measurable. The specific heat c_H is usually measured in zero field ($H = 0$), so this quantity will be calculated.

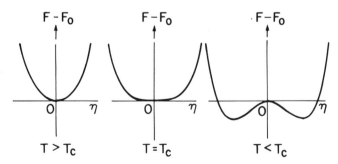

Figure 2.1. The Helmholtz free energy F plotted as a function of the order parameter η. At and above the critical temperature the minimum value of F corresponds to $\eta = 0$; below the critical temperature the minima corresponds to nonzero values of η.

Standard thermodynamics shows that c_H is given by

$$c_H = -T \frac{\partial^2 G}{\partial T^2}\bigg|_H$$

where $G(T, H)$ is the Gibbs free energy. The Gibbs and Helmholtz free energies are related by

$$G = F - HM$$

For small applied fields H we can substitute for M and F from Eq. 2.3 and Eqs. 2.4 or 2.5. This gives

$$c_H = -T \frac{d^2 F_0}{dT^2} \qquad (T > T_c) \tag{2.6}$$

and

$$c_H = -T \frac{d^2 F_0}{dT^2} - \frac{\alpha_0^2}{2\alpha_4} \qquad (T < T_c) \tag{2.7}$$

where it is assumed that the temperature dependence of $\alpha_4(T)$ can be neglected. The specific heat has a downward discontinuity of $\alpha_0^2/(2\alpha_4)$ at the critical temperature, as shown in Figure 2.2.

Experiments usually give a very different result for the specific heat from that predicted by the Ginzberg–Landau theory. The specific heat diverges at T_c in what is usually referred to as a *lambda anomaly*; this is shown by the broken line in Figure 2.2. Lambda anomalies correspond to rather weak divergences at the critical temperature, with a form similar to that of a logarithmic divergence (though not necessarily exactly a

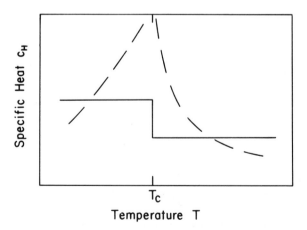

Figure 2.2. The variation of the specific heat c_H with temperature near a critical phase transition. The solid line shows the predictions of Ginzburg–Landau theory that there will be a step discontinuity downwards at T_c as the temperature is raised. Experimentally, the specific heat is found to diverge at T_c as shown by the dashed line. The divergence is close to logarithmic and is commonly referred to as a *lambda anomaly*.

logarithmic divergence). This result shows quite clearly that the Ginzburg–landau theory is inadequate.

We can also calculate the susceptibility for the Ginzburg–Landau theory. Measurements usually correspond to the isothermal susceptibility χ_T, which is given by

$$\chi_T^{-1} = \left.\frac{\partial^2 F}{\partial M^2}\right|_T = 2\alpha_2(T) + 12\eta^2 \alpha_4(T)$$

For temperatures above T_c, where $\eta = 0$, this yields

$$\chi_T = \frac{1}{2\alpha_0} \frac{1}{T - T_c} \qquad (T > T_c) \tag{2.8}$$

This is in fact the Curie–Weiss law with $(2\alpha_0)^{-1}$ equal to the Curie constant. It is found experimentally to be not too bad at temperatures well above T_c but to break down close to T_c. This derivation would predict just the opposite behavior, since a Taylor expansion about the critical point should be most accurate close to that point.

Although the Ginzburg–Landau theory has been shown to be solvable, it has so far not looked very useful, as it makes predictions that do not check out with experiment. In fact there are two types of continuous phase transition for which the Ginzburg–Landau theory seems to be satisfactory. These are for ferroelectrics and for type I superconductors. The common feature of these two cases is that the microscopic interactions are of long range, and the Ginzburg–Landau theory actually gives correct predictions for interactions of infinitely long range. However, for most critical phase transitions the interactions are short-ranged and it is necessary to go on to develop a better theory.

Suggested Further Reading

Landau and Lifshitz (1969)
Stanley (1971)

3

CRITICAL EXPONENTS

3.1. The Basic Idea

When a theory fails, as does the Ginzburg–Landau theory, it is often appropriate to try to rationalize all the experimental data phenomenologically. In the magnetic case, for example, there are measurements available near the critical point of the susceptibility, the specific heat, the correlation length, and the magnetization. The first three of these properties all become very large as the critical point is approached. Experiment cannot give a result of infinity for physical quantities, but the "better" the conditions the larger is the value measured, and it is generally believed that these quantities are in fact becoming infinite at the critical point. In an analogous way the magnetization tends to zero as the critical point is approached, so that the reciprocal of the magnetization tends to infinity.

Many careful measurements have been made of how these parameters diverge as the critical point is approached, and almost all of them seem to obey a simple power law. To give an example, if we define a reduced temperature t by

$$t = \frac{T - T_c}{T_c} \tag{3.1}$$

then, close to the critical point, measurements of the isothermal susceptibility χ_T, in small fields at temperatures just above T_c, fit the law

$$\chi_T = at^{-\gamma} \qquad (T > T_c) \tag{3.2}$$

where a and γ are constants.

This is illustrated in Figure 3.1, which shows measurements by Noakes et al. (1966) of the susceptibility of iron alloyed with 0.5% tungsten plotted against $T - T_c$. The plot is on a log–log scale and shows a straight line with slope -1.33, appropriate to $\gamma = 1.33$. The indices of the power laws, such as γ, are known as *critical exponents*. Their values vary somewhat from case to case, but the power law itself almost always holds. Table 3.1 gives a summary of these exponents for the properties that have been mentioned earlier. For each, the predictions of the Ginzburg–Landau theory are given and compared with the range of values found experimentally. It is clear that the Ginzburg–Landau theory is not quantitatively adequate. Since the theory would seem to depend only on the assumption that the free energy can be expanded as a Taylor

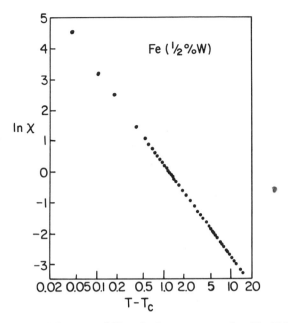

Figure 3.1. Log–log plot of the susceptibility of a ferromagnet against $T - T_c$ from Noakes et al. (1966). The plot is linear over almost three decades of $T - T_c$ with slope $\gamma = -1.333 \pm 0.015$, showing that in the critical region the susceptibility diverges at $(T - T_c)^{-\gamma}$.

series about the critical point, it must be concluded that the free energy is not an analytic function at the critical point, notwithstanding the fact that there is no reason to believe that the microscopic interactions are anything other than simple well-behaved functions. Such nonanalytic behavior can only be generated from simple microscopic interactions by taking an infinitely large system. This gives a warning that exact (or numerical) solutions of model systems of small size may not shed much light on continuous phase transitions.

Table 3.1. Definitions of Some Magnetic Critical Exponents, and Comparison between Predicted Values from the Ginzburg–Landau Theory and Range of Values Found Experimentally

Property	Critical Exponent	Power Law	Conditions	Ginzburg–Landau Prediction	Experiment
Susceptibility, χ_T	γ	$t^{-\gamma}$	$T > T_c$, $H = 0$	1	1.3–1.4
Susceptibility, χ_T	γ'	$(-t)^{-\gamma'}$	$T < T_c$, $H = 0$	1	—
Magnetization, M	β	$(-t)^{\beta}$	$T < T_c$, $H = 0$	0.5	0.2–0.4
Magnetization, M	δ	$H^{1/\delta}$	$T = T_c$,	3	3–6
Specific heat, c_H	α	$t^{-\alpha}$	$T > T_c$, $H = 0$	Discontinuous	−0.3–0.3
Specific heat, c_H	α'	$(-t)^{-\alpha'}$	$T < T_c$, $H = 0$	Discontinuous	−0.3–0.3
Correlation length ξ	ν	$t^{-\nu}$	$T > T_c$, $H = 0$	0.5	0.6–0.7
Correlation length, ξ	ν'	$(-t)^{-\nu'}$	$T < T_c$, $H = 0$	—	0.6–0.7

We will now take space to define critical exponents more carefully and to discuss their practical determination. If we say that a function $f(x)$ has critical exponent λ close to the critical point $x = 0$ as the critical point is approached from positive x, we mean

$$f(x) \sim x^\lambda \quad \text{when} \quad x \to 0+$$

The exponent λ can be defined by

$$\lambda = \lim_{x \to 0+} \frac{\ln f(x)}{\ln x}$$

so that λ is defined by the asymptotic behavior close to the critical point. If the critical point can be approached from negative x also, then we can analogously define a critical exponent λ' for $x \to 0-$.

In a real case there will be *correction terms* to this asymptotic law, such as

$$f(x) = Ax^\lambda(1 + a_y x^y + \cdots) \tag{3.3}$$

where y is greater than zero, so that the correction term dies away in the limit as x goes to zero. These correction terms limit the range of x over which the simple power behavior holds and reduce the accuracy with which critical exponents can be determined experimentally.

The standard experimental practice in dealing with critical properties is to plot the logarithm of $f(x)$ against the logarithm of x. An example of this was seen in Figure 3.1, where $f(x) = \chi$ and $x = T - T_c$ and these are plotted logarithmically. An immediate problem with making such a plot is that T_c must be known a priori, and this is often not the case. Uncertainties in T_c affect the plot very markedly at small x but have little effect at large x, and often the plot has to be made with T_c as a fitted parameter—fitted so as to keep the plot a straight line at the smallest values of $T - T_c$. Such a procedure clearly introduces experimental error into the determination of critical exponents. It becomes hard to estimate the uncertainty in the critical exponent, since the slope of the plot at small x is sensitive to the fitted value of T_c and the slope at large x is sensitive to correction terms. In consequence, practical determinations of critical exponents usually have limited accuracy and very few measurements are able to determine critical exponents to much higher absolute accuracy than 0.005; most experimental data show uncertainties in critical exponents in the region 0.01–0.02 because of the difficulties that have been outlined.

3.2. Inequality Relationships between Critical Exponents

It is possible to derive from thermodynamic arguments certain exact relationships between response functions. For example, in the magnetic case it can be shown that

$$\chi_T(c_H - c_M) = T\left(\frac{\partial M}{\partial T}\bigg|_H\right)^2 \tag{3.4}$$

and for a stable system it is necessary that χ_T, c_H, and c_M not be less than zero, so that

$$\chi_T c_H \geq T\left(\frac{\partial M}{\partial T}\bigg|_H\right)^2 \tag{3.5}$$

Near, but just below, the critical temperature in zero field, we expect that

$$\chi_T \sim (-t)^{-\gamma'} \tag{3.6}$$

$$c_H \sim (-t)^{-\alpha'} \tag{3.7}$$

$$\frac{\partial M}{\partial T}\bigg|_H \sim (-t)^{\beta-1}$$

so that

$$A(-t)^{-(\alpha'+\gamma')} \geq B(-t)^{2(\beta-1)}$$

where A and B are positive constants. The term in the temperature T in Eqs. 3.4 or 3.5 does not diverge at $T = T_c$ and so, sufficiently close to T_c, can be replaced with by constant:

$$\ln A - (\alpha' + \gamma') \ln(-t) \geq \ln B + 2(\beta - 1)\ln(-t)$$

In the limit as the critical temperature is approached from below, $\ln(-t)$ tends to negative infinity and the constant terms $\ln A$ and $\ln B$ can be dropped, so that

$$-(\alpha' + \gamma') \leq 2(\beta - 1), \qquad \alpha + 2\beta + \gamma' \geq 2 \tag{3.8}$$

This is known as the *Rushbrooke inequality* (Rushbrooke 1963). For the Ginzburg–Landau model, $\alpha' = 0$, $\beta = \frac{1}{2}$, and $\gamma' = 1$, so the inequality becomes an equality. At first sight it may seem surprising that this should happen, because the inequality was generated by dropping the term involving c_M. However, on closer examination it becomes apparent that if c_M diverges more slowly than c_H the inequality will become an equality in the limit; furthermore, c_M cannot diverge more rapidly than c_H else the left-hand side of the equation becomes negative and the right-hand side must be positive. Only if c_H diverges with the same critical exponent as c_M is there a possibility of the equation yielding an inequality rather than an equality. Griffiths (1965) lists 17 such inequalities that can be derived with varying degrees of rigor.

Suggested Further Reading

Fisher (1967)
Griffiths (1965)
Stanley (1971)

4

UNIVERSALITY, STANDARD MODELS, AND SOLVABLE MODELS

4.1. Universality

A look at all the experimental data on critical exponents reveals certain apparently systematic trends. In the (relatively few) cases where critical exponents can be calculated with reasonable accuracy from theoretical models, the same features are apparent. These are usually expressed as the hypothesis of *universality* (Griffiths 1970, Griffiths and Wheeler 1970, Kadanoff 1971) which states that:

For a continuous phase transition the static critical exponents depend on the following three properties and nothing else:

1. The dimensionality of the system, d
2. The dimensionality of the order parameter, D (or more precisely, the symmetry of the order parameter; in simple cases this is equivalent to the number of dimensions in which the order parameter is free to vary)
3. Whether the forces are of short or long range

This represents an enormous generalization, because it implies that the nature of the microscopic interaction is irrelevant (except for its range, if that range is long). It is also irrelevant whether the system is quantum-mechanical or classical. For continuous transitions within the solid state, such as magnetic transitions, the critical exponents are predicted to be the same whatever the crystal structure.

The basic idea is not new; it was anticipated in the nineteenth century by Van der Waals in formulating the law of *corresponding state,* which says that all gases and liquids have the same equation of state if the variables for pressure, volume, and temperature are replaced by dimensionless "reduced" variables, which are the ratio of the value of the variable to its value at the critical point. This law has been found to work quite well except for polar molecules, for which long range forces are present.

Of course, the modern form of universality is a considerable generalization of Van der Waals' ideas. It has the status of a hypothesis, since it has not been "proved" from more basic ideas, and it must be judged by reference to experiment, where in fact measurements seem to support the hypothesis. There are still doubts as to how good universality is (for instance, Haldane (1983) has conjectured different behavior of one-dimensional magnetic systems with integral and half-integral spins) but, nonetheless, it forms a reasonable working hypothesis from which to proceed to other questions. In the next two chapters we will develop some insight as to why universality might work as well as it does.

In addition to the static properties covered by universality, there are also dynamic properties that show divergences at critical points that obey power laws with critical exponents. Universality does not work for these exponents. A simple example of the breakdown is provided by dynamics of three-dimensional Heisenberg systems. Ferromagnets have long-wavelength excitations in the ordered state (spin waves) that have energy proportional to q^2, where q is the wavevector, while antiferromagnets have long-wavelength excitations that have energy proportional to q. This results in different dynamic critical properties for systems that differ only in the sign of the coupling between spins. This difference can be expressed by the fact that for the ferromagnet the order parameter (the magnetization) commutes with the Hamiltonian and so is conserved, while for the antiferromagnet the order parameter is the staggered magnetization, which does not commute with the Hamiltonian and is not conserved. Halperin et al. (1972, 1974) proposed that universality holds for dynamic properties if a fourth condition is added to the three set out above. This is:

4. The dynamic critical exponents also depend on the conservation laws of the system.

4.2. Standard Models

Universality offers the prospect of a simplification in the theory of continuous phase transitions. As we have seen, there are many very different systems that show such phase transitions: for most of these the microscopic Hamiltonian is quite complicated and there is little hope of making direct calculations near the critical point. What universality does is to enable us to choose the simplest theoretical model that we can think of for any particular *universality class* (i.e., any particular values of D, d, and the range of forces). If we can solve for the critical properties of this model, then the result will apply to all models within the universality class. These "simplest" models are magnetic, which is why critical phenomena are often discussed in the language of magnetism. The models are for different values of the dimensionality of the order parameter and for short-range forces confined to neighboring sites (long-range forces usually give the Ginzburg–Landau model). It is convenient to define the following four *standard models* in addition to the Ginzburg–Landau model

(1) The *Ising model* corresponding to $D = 1$ with Hamiltonian $\mathcal{H}$ given

by

$$\mathcal{H} = -\sum_{\mathbf{n}} \sum_{\mathbf{i}}' J_{\mathbf{i}} S_{\mathbf{n}}^z S_{\mathbf{n+i}}^z \qquad (4.1)$$

where $S_{\mathbf{n}}^z$ is the z component of spin on the site at $\mathbf{n}$. The exchange parameter $J_{\mathbf{i}}$ couples spins on sites at $\mathbf{n}$ and at $\mathbf{n+i}$ where $\mathbf{i}$ is a nearest-neighbor vector. The second sum is given a prime to indicate that the sum over $\mathbf{i}$ is to be restricted so that each pair of interacting spins is only included once. The order parameter for the ferromagnetic case ($J > 0$) is $\sum_{\mathbf{n}} S_{\mathbf{n}}^z$, which is one-dimensional. This universality class is appropriate for Ising magnets, liquid–gas transitions, ordering in alloys, and mixing in liquids.

(2) The *X–Y model* corresponding to $D = 2$ with Hamiltonian $\mathcal{H}$ given by

$$\mathcal{H} = -\sum_{\mathbf{n}} \sum_{\mathbf{i}}' J_{\mathbf{i}} (S_{\mathbf{n}}^x S_{\mathbf{n+i}}^x + S_{\mathbf{n}}^y S_{\mathbf{n+i}}^y) \qquad (4.2)$$

Here the spins have two components (x and y) and the order parameter is the vector sum of these spins, which is two-dimensional. This universality class applied to "easy-plane" magnets and to superfluidity in liquid ^{4}He.

(3) The *Heisenberg model* corresponding to $D = 3$ with Hamiltonian

$$\mathcal{H} = -\sum_{\mathbf{n}} \sum_{\mathbf{i}}' J_{\mathbf{i}} (S_{\mathbf{n}}^x S_{\mathbf{n+i}}^x + S_{\mathbf{n}}^y S_{\mathbf{n+i}}^y + S_{\mathbf{n}}^z S_{\mathbf{n+i}}^z) \qquad (4.3)$$

Here the spin is a three-dimensional vector. The model applies to isotropic magnetic materials.

(4) The *spherical model* corresponding to $D = \infty$. This model assumes the spin $S_{\mathbf{n}}$ has an infinite number of dimensions. It does not seem to correspond to any actual system, but it has the attraction of being solvable exactly (Berlin and Kac 1952, Stanley 1968).

It might be added that these standard models by no means exhaust the universality classes actually found, since magnetic systems are known that correspond to $D = 4, 6, 8$, and 12 (Mukamel and Krinsky 1976, Bak and Mukamel 1976), and to the rotation group of symmetry SO(3) (Kawamura 1986). For example, erbium antimonide, ErSb, has a simple sodium chloride structure. At low temperatures an antiferromagnetic structure is formed that consists of ferromagnetic sheets in (111) planes (Shapiro and Bak 1975). Anisotropy confines the moments in any sheet to within the (111) plane, so that its effective dimensionality is 2; there are, however, four equivalent (111) directions in a cubic material, so eight components are needed to describe the ordering.

4.3. Solvable Models

It is an unfortunate fact that, although a number of exact solutions are available for static critical properties, almost all the critical transitions that occur in the real world correspond to universality classes that cannot be solved exactly. It is perhaps useful, however, to list some universality classes within which models have been solved:

1. All cases in one dimension ($d = 1$). Unfortunately these do not show continuous phase transitions and are of limited use for our present purpose.
2. All cases in four or more dimensions ($d \geq 4$). These give the Ginzburg–Landau solution, but do not correspond to any physical reality.
3. The Ising model in two dimensions ($d = 2$, $D = 1$). This is Onsager's famous solution (Onsager 1944), which gives a continuous phase transition and represents the earliest major progress beyond the Ginzburg–Landau model.
4. The spherical model ($D = \infty$) in any number d of dimensions (Berlin and Kac 1952, Stanley 1968).
5. All cases in which the range of the interactions is infinite; these follow the Ginzburg–Landau model.

The exact solutions of Onsager and of Berlin and Kac both give power law variations of response functions near the critical point and serve to reinforce belief in the idea of characterizing continuous phase transitions by critical exponents. Most actual critical systems correspond to the three-dimensional case ($d = 3$) and it is an unfortunate fact that the commonly occurring universality classes in three dimensions have not been solved exactly.

Suggested Further Reading

Kadanoff (1971)

5

SCALING

5.1. The Scaling Approach

Although universality has simplified the problem, the direct approach of solving the standard models for commonly occuring universality classes is still too difficult. In this chapter a less rigorous approach, due to Kadanoff (1966), will be introduced using the magnetic language of the standard models.

In the critical region (that is, near the critical point) there will be large volumes, with dimensions of the order of ξ, where the magnetization density is fairly constant, and other volumes of the same sort of size also with fairly constant magnetization density but with the magnetization in different directions. These regions only fluctuate slowly in time and, since ξ becomes infinite at the critical point, the volume of the regions is very large near the critical point.

Suppose a correlated region is divided into cells with L lattice sites on each side. In a d-dimensional lattice this will give L^d sites per cell. L is assumed to be large, but small compared with ξ/a, where a is the lattice parameter. This condition can always be met near the critical point, since ξ becomes infinite at the critical point.

Each cell will have a Hamiltonian that expresses its interaction with the neighboring cells. We might expect this Hamiltonian to show only short-range interactions to neighboring cells and this Hamiltonian might be expected to be expressible in terms of the parameters S and H_e, where S is the effective cell spin and H_e is the effective field between cells. Since we are in a correlated region ($L \ll \xi/a$), S and H_e should be similar for neighboring cells, and H_e should be uniform throughout a cell. Both scaling theory, and later the renormalization group theory, will make the assumption that the form of the Hamiltonian does not change as we vary the cell size; all that does change is the parameters (H_e and S) that go into the Hamiltonian.

At this point it is convenient to go over to dimensionless variables. The length variables L and ξ/a are already dimensionless and we replace H_e and T by the dimensionless variables

$$t = \frac{T - T_c}{T_c}, \quad h = \frac{g\mu_B H_e}{k_B T} \tag{5.1}$$

where μ_B is the Bohr magneton and k_B is Boltzmann's constant.

The Gibbs free energy of a cell, $G(T, H)$, can be written in terms of t

and h as $G(t, h)$. Now let us consider what happens if we change the length scale L. Scaling theory argues that the functional form of the free energy does not change with L, only the parameters t and h. Suppose we change the cell size from La to lLa, then the parameters of G change from t and h to $\tilde{t}$ and $\tilde{h}$ (t varies because T_c depends on cell spin and exchange parameters, which in turn depend on L), and the free energy changes to $G(\tilde{t}, \tilde{h})$. We also note that free energy is an extensive property so that

$$l^d G(t, h) = G(\tilde{t}, \tilde{h}) \tag{5.2}$$

Now we must relate t to $\tilde{t}$ and h to $\tilde{h}$. As h is small, we would expect a linear relationship between h and $\tilde{h}$ and we assume a linear relationship between t and $\tilde{t}$. Given these assumptions it can be proved (Cooper 1968) that

$$\tilde{h} = l^x h \tag{5.3}$$

and

$$\tilde{t} = l^y t$$

This gives

$$l^d G(t, h) = G(l^y t, l^x h) \tag{5.4}$$

This relation says mathematically that G is a generalized homogeneous function, a mathematical form that was put forward independently by Widom (1965) and by Domb and Hunter (1965) before Kadanoff's derivation. We have followed Kadanoff's derivation because it uses ideas that are also germane to the renormalization group theory that will be the subject of the next chapter.

We can derive a second homogeneous functional relationship, this time for the correlation length ξ. We expect ξ to be a function only of the temperature t and field h:

$$\xi = \xi(t, h)$$

Since we are using reduced, or dimensionless, units, the length scale for the dimensionless quantity ξ/a must be the size of our cell, La, and in absolute units the correlation length is $L\xi(t, h)$. Now if we change our cell size to lL, with concomitant changes of t and h to $\tilde{t}$ and $\tilde{h}$, the correlation length becomes $lL\xi(\tilde{t}, \tilde{h})$. In fact, the correlation length must be independent of the particular cell size that we choose, so

$$L\xi(t, h) = lL\xi(\tilde{t}, \tilde{h})$$

It follows that

$$l^{-1}\xi(t, h) = \xi(l^y t, l^x h) \tag{5.5}$$

which gives another generalized homogeneous function.

5.2. Solution of Scaling Equations: Scaling Laws

Let us now solve the two Eqs. (5.4 and 5.5) in zero magnetic field, when they become

$$l^d G(t) = G(l^y t) \qquad (5.6)$$

and

$$l^{-1} \xi(t) = \xi(l^y t) \qquad (5.7)$$

These equations should be valid for any large positive value of l, which will include the value at which

$$l^y t = 1$$

since we can make t arbitrarily small. We have assumed t to be positive, so that the temperature is just above T_c. At this value of l,

$$G(t) = t^{d/y} G(1) \qquad (5.8)$$

and

$$\xi(t) = t^{-1/y} \xi(1) \qquad (5.9)$$

Table 3.1 shows that just above T_c

$$\xi(t) \sim t^{-\nu} \qquad (5.10)$$

Comparison between Eqs. 5.9 and 5.10 shows that

$$y = \nu^{-1} \qquad (5.11)$$

We can use Eq. 5.8 for the Gibbs free energy to calculate the specific heat in constant field (c_H) at $H = 0$ since

$$c_H = -T \frac{\partial^2 G}{\partial T^2}\bigg|_H$$

so that

$$c_H = -T \frac{d}{y}\left(\frac{d}{y} - 1\right) t^{d/y - 2} G(1) T_c^{-2}$$

and, from Table 3.1, just above T_c

$$c_H \sim t^{-\alpha}$$

Very close to T_c, T can be regarded as a constant, so that

$$d/y - 2 = -\alpha, \qquad 2 - \alpha = \frac{d}{y} = d\nu \qquad (5.12)$$

An equality like this between critical indices is known as a *scaling law*. This particular equality is not obeyed by the Ginzburg–Landau theory in three dimensions, where $\alpha = 0$ and $\nu = 0.5$; however, it is a very promising sign that it is obeyed by the exact solution for the Ising model in two dimensions ($\alpha = 0$ and $\nu = 1$) and for the spherical model in three dimensions ($\alpha = -1$ and $\nu = 1$). There is also an inequality between critical exponents (Josephson 1967) that requires

$$d\nu \geq 2 - \alpha$$

This, like several other of the scaling laws, corresponds to replacing a more-or-less rigorous inequality with an equality.

Previously, we took the temperature to be just above T_c ($t > 0$), but we could have used the same arguments for temperatures just below T_c ($t < 0$) and have set

$$l^y(-t) = 1$$

Then it would follow that

$$G(t) = (-t)^{d/y} G(-1), \qquad \xi(t) = (-t)^{-1/y} \xi(-1)$$

while reference to Table 3.1 shows that

$$\xi(t) \sim (-t)^{-\nu'}$$

and

$$c_H \sim (-t)^{-\alpha'}$$

so that

$$\nu' = \nu \qquad \text{and} \qquad \alpha' = \alpha \tag{5.13}$$

This equality of primed and unprimed critical exponents is a general feature of scaling theory.

We now solve the scaling equations for the magnetization in a small field. Thermodynamics gives the magnetization, $M(T, H)$, by

$$M(T, H) = -\left.\frac{\partial G(T, H)}{\partial H}\right|_T \tag{5.14}$$

We get the scaling transformation of the magnetization by differentiating the equation that gives the transformation for the free energy (Eq. 5.4):

$$l^d \frac{\partial G(t, h)}{\partial h} = \frac{\partial G(\tilde{t}, \tilde{h})}{\partial h} = l^x \frac{\partial G(\tilde{t}, \tilde{h})}{\partial \tilde{h}} \tag{5.15}$$

It follows that in zero magnetic field

$$l^d M(t, 0) = l^x M(\tilde{t}, 0) = l^x M(l^y t, 0)$$

This holds for all positive values of l, and as before we put

$$l^y(-t) = 1$$

for a temperature just below T_c, so that t is negative. Then

$$M(t, 0) = (-t)^{(d-x)/y} M(-1, 0)$$

and from Table 3.1

$$M(t, 0) \sim (-t)^\beta$$

so that
$$\beta = \frac{d-x}{y}$$

Combining this with Eq. 5.11 gives
$$\beta = (d-x)v \tag{5.16}$$

Equations 5.14 and 5.15 can also be used to derive the variation of the magnetization as a function of H. This gives
$$l^d M(t, h) = l^x M(\tilde{t}, \tilde{h}) = l^x M(l^y t, l^x h) \tag{5.17}$$

and putting $l^x h = 1$ at the critical temperature ($t = 0$) leads to
$$M(0, h) = h^{(d-x)/x} M(0, 1)$$

Reference to Table 3.1 shows that
$$M(0, h) \sim h^{1/\delta}$$

so
$$\frac{1}{\delta} = \frac{d-x}{x}$$

or
$$x = (d-x)\delta \tag{5.18}$$

The isothermal susceptibility can be calculated from the magnetization by means of the relationship
$$\chi_T = \frac{\partial M}{\partial H}\bigg|_T$$

Differentiation of Eq. 5.17 yields
$$\frac{\partial M(t, h)}{\partial h}\bigg|_t = l^{x-d}\frac{\partial M(\tilde{t}, \tilde{h})}{\partial h}\bigg|_t = l^{2x-d}\frac{\partial M(\tilde{t}, \tilde{h})}{\partial \tilde{h}}\bigg|_t$$

In zero field ($h = 0$) this gives
$$\chi_T = l^{2x-d}\frac{\partial M(l^y t, \tilde{h})}{\partial \tilde{h}}\bigg|_t\bigg|_{\tilde{h}=0}$$

For temperatures just below T_c ($t < 0$) we put
$$l^y(-t) = 1$$

to obtain
$$\chi_T \equiv (-t)^{-(2x-d)/y}\frac{dM(-1, \tilde{h})}{d\tilde{h}}\bigg|_{\tilde{h}=0}$$

Reference to Eq. 3.5 shows, that, at zero field,
$$\chi_T \sim (-t)^{-\gamma'}$$

so

$$\gamma' = \frac{2x-d}{y} = (2x-d)v \tag{5.19}$$

Combination of Eqs. 5.12, 5.13, 5.16, 5.18, and 5.19 yields two scaling laws:

$$\gamma' = \beta(\delta - 1) \tag{5.20}$$

and

$$\gamma' + \alpha' + 2\beta = 2 \tag{5.21}$$

The second of these is the Rushbrooke inequality, which has already been derived (Eq. 3.6) with the inequality replaced by an equality.

5.3. Correlation Function Approach

The scattering properties of a system depend on correlation functions, and critical scattering can be used to measure the correlation function in the critical region. Thus, it will be pertinent to figure out the scaling properties of correlation functions.

Let us assume, following the standard models, that we have spins **S** of dimensionality **D** lying on a lattice with sites **R** in d dimensions. Then the spin correlation function $C(\mathbf{R}, t, h)$ is defined by

$$\begin{aligned}C(\mathbf{R}, t, h) &= \langle \mathbf{S_0} \cdot \mathbf{S_R} \rangle - \langle \mathbf{S_0} \rangle \cdot \langle \mathbf{S_R} \rangle \\ &= \langle (\mathbf{S_0} - \langle \mathbf{S_0} \rangle) \cdot (\mathbf{S_R} - \langle \mathbf{S_R} \rangle) \rangle \end{aligned} \tag{5.22}$$

For most properties of interest, this function is averaged over all origin sites, **0**, and where appropriate it will be assumed implicitly that this average is to be taken. The same ideas that were used for the scaling of the free energy and of the correlation length can be applied to derive the correlation function between blocks of spins of size lLa. As before the superscript $\tilde{}$ refers to blocks of spins, and the block correlation function is given by

$$C(\tilde{\mathbf{R}}, \tilde{t}, \tilde{h}) = \langle (\tilde{\mathbf{S}}_0 - \langle \tilde{\mathbf{S}}_0 \rangle) \cdot (\tilde{\mathbf{S}}_{\tilde{\mathbf{R}}} - \langle \tilde{\mathbf{S}}_{\tilde{\mathbf{R}}} \rangle) \rangle$$

The same line of reasoning that leads to Eq. 5.5 gives the relation

$$\tilde{\mathbf{R}} = l^{-1}\mathbf{R}$$

if **R** is in dimensionless units and the block size is used as the length scale.

The Hamiltonian must be the same whether the system is divided into blocks of size Ll or size L, so the term in the field must be the same in the two cases. This implies that

$$h \sum_{\mathbf{R}} \mathbf{S_R} = \tilde{h} \sum_{\tilde{\mathbf{R}}} \tilde{\mathbf{S}}_{\tilde{\mathbf{R}}} \tag{5.23}$$

If there are N blocks of size L, there are Nl^{-d} blocks of size lL and

$$\langle \mathbf{S} \rangle = N^{-1} \sum_{\mathbf{R}} \mathbf{S_R}$$

$$\langle \tilde{\mathbf{S}} \rangle = N^{-1} l^d \sum_{\tilde{\mathbf{R}}} \tilde{\mathbf{S}}_{\tilde{\mathbf{R}}}$$

Putting these into Eq. 5.23, and remembering Eq. 5.3, gives

$$h \langle S \rangle = \tilde{h} l^{-d} \langle \tilde{S} \rangle$$
$$\langle S \rangle = l^{x-d} \langle \tilde{S} \rangle$$

The correlation function now becomes

$$C(\mathbf{R}, t, h) = l^{2(x-d)} C(\tilde{\mathbf{R}}, \tilde{t}, \tilde{h})$$
$$= l^{2(x-d)} C(l^{-1}\mathbf{R}, l^y t, l^x h) \quad (5.24)$$

This is again a generalized homogeneous function, but with three variables rather than with two variables as encountered before. The same method of solution can be used. We consider just the zero-field case and put

$$l^{-1} |\mathbf{R}| = 1$$

to get

$$C(\mathbf{R}, t, 0) = |\mathbf{R}|^{2(x-d)} C(1, |\mathbf{R}|^y t, 0)$$

Now from Eqs. 5.10 and 5.11,

$$|\mathbf{R}|^y t = (|\mathbf{R}| t^{1/y})^y$$
$$\sim (|\mathbf{R}|/\xi)^y$$

so that

$$C(\mathbf{R}, t, 0) = |\mathbf{R}|^{2(x-d)} f(|\mathbf{R}|/\xi) \quad (5.25)$$

This is the general form of the correlation function in zero field predicted by the scaling laws. It is conventional to express the correlation function at the critical point by the equation

$$C(\mathbf{R}, 0, 0) \sim |\mathbf{R}|^{2-d-\eta} \quad (5.26)$$

where η is a critical exponent defined by this equation. It follows that

$$2(x - d) = 2 - d - \eta \quad (5.27)$$

Combination with Eq. 5.19 gives the scaling law

$$(2 - \eta)\nu = \gamma \quad (5.28)$$

We might add that it is expected that the scaling function f in Eq. 5.25 will include an exponential term $\exp(-|\mathbf{R}|/\xi)$ and, at a fixed temperature above T_c, the correlation function will have the form (Ornstein and Zernike 1914, Fisher 1964, Fisher and Burford 1967).

$$C(\mathbf{R}, t, 0) \sim |\mathbf{R}|^{(1-d)/2} \xi^{(3-d)/2-\eta} \exp(-|\mathbf{R}|/\xi) \quad (5.29)$$

Scaling

For many applications it is more natural to use the correlation function formalism in reciprocal space rather than in real space. The correlation function in reciprocal space, $\hat{C}(\mathbf{q}, t, h)$ is defined as the Fourier transform, in d dimensions, of the real-space function

$$\hat{C}(\mathbf{q}, t, h) = \sum_{\mathbf{R}} e^{i\mathbf{q} \cdot \mathbf{R}} C(\mathbf{R}, t, h)$$

$$= \sum_{\mathbf{R}} e^{i\mathbf{q} \cdot \mathbf{R}} [\langle \mathbf{S}_0 \cdot \mathbf{S}_\mathbf{R} \rangle - \langle \mathbf{S}_0 \rangle \langle \mathbf{S}_\mathbf{R} \rangle] \quad (5.30)$$

The last term in this equation is zero except at reciprocal lattice points. It will be dropped, with the understanding that the equation now refers only to values of $\mathbf{q}$ that are not reciprocal lattice points.

As well as defining the correlation function in reciprocal space, it is useful to define spin variables $\mathbf{S}_\mathbf{q}$ in reciprocal space as the Fourier transforms of the corresponding real-space spin variables. Let

$$\mathbf{S}_\mathbf{R} = N^{-1/2} \int d\mathbf{q}\, e^{-i\mathbf{q} \cdot \mathbf{R}} \mathbf{S}_\mathbf{q} \quad (5.31)$$

where the lattice has N sites and the integral is over the first Brillouin zone (in d dimensions). Then

$$\hat{C}(\mathbf{q}, t, h) = N^{-1} \sum_{\mathbf{R}} e^{i\mathbf{q} \cdot \mathbf{R}} \left\langle \int\int d\mathbf{q}'\, d\mathbf{q}''\, e^{-i\mathbf{q}'' \cdot \mathbf{R}} \mathbf{S}_{\mathbf{q}'} \cdot \mathbf{S}_{\mathbf{q}''} \right\rangle$$

Since

$$\sum_{\mathbf{R}} e^{i\mathbf{q} \cdot \mathbf{R}} e^{-i\mathbf{q}'' \cdot \mathbf{R}} = N\delta(\mathbf{q} - \mathbf{q}'')$$

the equation simplifies to become

$$\hat{C}(\mathbf{q}, t, h) = \int d\mathbf{q}' \langle \mathbf{S}_{\mathbf{q}'} \cdot \mathbf{S}_\mathbf{q} \rangle \quad (5.32)$$

We shall make use of this equation in the next chapter. Meanwhile, we can put the scaling form of the real-space correlation function (Eq. 5.25) into Eq. 5.30 for the reciprocal-space correlation function to get (Eq. 5.27 is also used)

$$\hat{C}(\mathbf{q}, t, 0) = |q|^{\eta-2} \hat{f}(|q|\xi) \quad (5.33)$$

This form can actually be verified experimentally by scattering experiments that measure reciprocal-space correlation functions. At the critical temperature ($t = 0; \xi = \infty$) this becomes

$$\hat{C}(\mathbf{q}, 0, 0) = |q|^{\eta-2} \hat{f}(0) \quad (5.34)$$

Away from the critical temperature, the Fourier transform of the correlation function given by Eq. 5.29 leads to complicated analytic forms (Fisher and Burford 1967, Ritchie and Fisher 1972). In three dimensions, it is a fair approximation to put $\eta = 0$ to obtain the Lorentzian form originally proposed by Ornstein and Zernike (1914)

$$\hat{C}(\mathbf{q}, t, 0) \sim \frac{1}{\kappa_1^2 + q^2} \tag{5.35}$$

with $\kappa_1 = \xi^{-1}$ and $d = 3$ and $\eta = 0$.

If one wishes to go beyond this and put in a nonzero value for η, a simple approximation is produced by noting that at $q = 0$ the correlation function is directly proportional to the susceptibility so that

$$\hat{C}(\mathbf{0}, t, 0) \sim \chi(t)$$
$$\sim t^{-\gamma} \tag{5.36}$$

Now $\kappa_1^2 \sim t^{2\nu}$ and the scaling law of Eq. 5.28 gives

$$\gamma = 2\nu(1 - \eta/2) \tag{5.37}$$

so that if the correlation function is written as

$$\hat{C}(\mathbf{q}, t, 0) \sim \frac{1}{(\kappa_1^2 + \psi q^2)^{1-\eta/2}} \tag{5.38}$$

then Eq. 5.37 is satisfied. Fisher (1964) has proposed this approximation as an improvement on that of Eq. 5.35 with $\psi = 1$. Later, Fisher and Burford (1967) suggest as a better approximation that

$$\psi = \frac{1}{1 - \frac{1}{2}\eta} \tag{5.39}$$

Another form for the correlation function used in literature is (Birgeneau et al. 1980).

$$\hat{C}(\mathbf{q}, t, 0) \sim \frac{\kappa_1^\eta}{\kappa_1^2 + q^2} \tag{5.40}$$

This satisfies Eq. 5.36.

5.4. General Features of Scaling Theory

Scaling theory expresses all the static critical exponents in terms of just two parameters, x and y. The theory gives no information about the values of x and y, so it does not predict critical exponents absolutely. The relationships between critical exponents are known as scaling laws: these can be classified effectively into three groups:

1. Scaling laws that put critical exponents above T_c equal to those below T_c (cf. Eq. 5.13).
2. Scaling laws that do not involve the dimensionality d (cf. Eqs. 5.20, 5.21, 5.29).

3. Scaling laws that involve the dimensionality d (cf. Eqs. 5.12, 5.16, 5.18, 5.19). These are known as the *hyperscaling laws* and follow from Eqs. 5.2 and 5.4, which give the scaling of the free energy.

Both the second and third of these categories correspond to replacing exact inequality relationships by equalities. The Ising model in two dimensions obeys all the scaling laws, but the Ginzburg–Landau theory only obeys the hyperscaling laws in four dimensions. For the spherical model, hyperscaling is satisfied if $d \leq 4$ but fails in more than four dimensions. These facts lead us to speculate that there is something special about a dimensionality of 4; we will see in the next chapter that this is indeed the case. Also, since the solution of the spherical model is exact, hyperscaling seems to fail in more than four dimensions! Experiment and model calculations seem to agree with hyperscaling for four or fewer dimensions. For a discussion of how hyperscaling can break down in more than four dimensions, the reader is referred to Fisher (1982, particularly appendix D).

In Table 5.1, approximate values of critical exponents are given for various models. For the Ginzburg–Landau, the spherical, and the two-dimensional Ising models the exponents are exact, while for the other models values are taken from approximate calculations of γ and ν and these are used to calculate the other exponents using the scaling laws. Although no scaling law involves the spin-dimensionality, D, it is apparent from the table that the critical exponents do change as D changes. The scaling parameter y $(= \nu^{-1})$ depends markedly on D, while x depends on weakly, if at all, on D.

Table 5.1. Approximate Values of Critical Exponents for Various Models[a]

Model	Ginzburg–Landau	Ising	Ising	X–Y	X–Y	Heisenberg	Spherical
D	any	1	1	2	2	3	∞
d	any	2	3	2	3	3	3
γ	1.0	1.75	1.2378 ±0.0006	—	1.316 ±0.009	1.388 ±0.003	1
ν	0.5	1	0.6312 ±0.0003	—	0.669 ±0.007	0.707 ±0.003	1
x	—	1.875	2.481 ±0.001	—	2.484 ±0.009	2.482 ±0.005	2.5
α	—	0	0.106	—	−0.01	−0.121	−1
β	0.5	0.125	0.326	—	0.345	0.367	0.5
δ	3	15	4.78	15	4.81	4.78	5
η	0	0.25	0.039	0.25	0.03	0.037	0

[a] The first two colums and the last column are exact. Values of γ and ν for the Ising model and the X–Y model in three dimensions are from George and Rehr (1984) and Baker et al. (1978), respectively. Values for the Heisenberg model in three dimensions are the weighted means of the values of γ and ν quoted in Table 1 of Ferer and Hamid-Aidinejad (1986). In these three cases, the remaining exponents are calculated using various scaling laws. Column 4 is from Kosterlitz (1974). Dashes indicate quantities that do not follow power laws in the critical region.

The final test of theory must be reference to experiment: here scaling stands up well, as it is found to hold within experimental error in almost every case. It might be added that, as we have seen, it is difficult to measure critical exponents with high accuracy, so that the comparison with experiment is often not as searching as might be wished.

The general success of scaling augurs well for the underlying hope, contained in universality, that there exists a general theory of continuous phase transitions.

Suggested Further Reading

Kadanoff (1966)
Stanley (1971)
Fisher (1982)

6

THE RENORMALIZATION GROUP

In the renormalization group approach we use the same general methods as in scaling theory, so that the scaling laws are retained, but go further in evaluating the block (renormalized) variables explicitly. The basic assumption is that changes in length scale merely alter the parameters of the Hamiltonian and not the underlying form of the Hamiltonian. Solutions that comply with the assumption can be found, with varying degrees of accuracy, for a wide variety of Hamiltonians. The development in this section follows along lines set out by Wilson and Kogut (1974).

Suppose that after n scaling transformations the system is a cube of side $l^n La$ with Hamiltonian $\mathcal{H}_n$. A further change of scale is made to give a cube of side $l^{n+1} La$ with Hamiltonian $\mathcal{H}_{n+1}$. There will be a transformation τ such that

$$\tau(\mathcal{H}_n) = \mathcal{H}_{n+1}$$

At the critical point, the correlation length ξ becomes infinitely large, and the intensive properties of the system should be unchanged by the transformation τ. Thus there should be a limiting function $\mathcal{H}^*$ such that

$$\tau(\mathcal{H}^*) = \mathcal{H}^*$$

$\mathcal{H}^*$ is called a *fixed point* of τ. The fixed point of τ should be independent of $\mathcal{H}_0$, although in general several fixed points may exist so that the system can exhibit several forms of cooperative behavior. It may be necessary to determine which (if any) fixed points result from the sequence starting with $\mathcal{H}_0$, and how this depends on the choice of $\mathcal{H}_0$.

6.1. The Gaussian Model

Take a standard Ising model (cf. Eq. 4.1) with Hamiltonian

$$\mathcal{H} = -\sum_n \sum_j{}' J_j S_n S_{n+j} \qquad (6.1)$$

where S_n is the z component of the spin on the nth site, restricted such that $S = \pm 1$. This is equivalent to the Ising model for quantum spins of $\pm \frac{1}{2}$. The partition function Z for this model is defined as

$$Z = \sum_{S_1=\pm 1} \sum_{S_2=\pm 1} \cdots \sum_{S_n=\pm 1} \cdots \exp(-\beta \mathcal{H})$$

where $\beta = 1/k_B T$ and the sum is over all possible spin configurations of the system, that is, over $S_n = \pm 1$ for all sites n. For any particular site n, the sum over $S_n = \pm 1$ can be replaced by an integral

$$\sum_{S_n = \pm 1} \rightarrow \int_{-\infty}^{\infty} dS_n\, \delta(S_n^2 - 1)$$

where δ is the Dirac delta function and the discrete function S_n has been replaced by a continuous variable. The partition function can be written as

$$Z = \prod_n \int_{-\infty}^{\infty} dS_n\, \delta(S_n^2 - 1) \exp(-\beta \mathcal{H}) \qquad (6.2)$$

In the renormalization group approach, the delta function in this equation is replaced by a continuous weighting function. This is done by setting

$$\delta(S_n^2 - 1) = \exp[W(S_n)] \qquad (6.3)$$

The weighting function $W(S_n)$ should become large and negative except near $S_n = \pm 1$. In practice we choose the most realistic function for which we can actually solve the problem! A possibility is

$$W(S_n) = -bS_n^2/2 - uS_n^4 \qquad (6.4)$$

This leads to weightings most like the original delta function when $u = -b/4$ ($u > 0$), and in fact the original delta function is recovered exactly here in the limit as u tends to infinity.

If u is set equal to zero we have the *Gaussian model,* so called because the delta function is replaced by a Gaussian function. This model has the virtue of being exactly solvable; it was first solved by Berlin and Kac (1952) and the solution can be mapped onto the solution for the spherical model. The more general weighting function given above is not a solved case, but solutions can be derived in perturbation theory if u can be treated as a small perturbation of the Gaussian model.

We will use a method of solution for the Gaussian model that involves changing the block sizes in reciprocal space, rather than the real-space method developed for scaling theory in the previous section. This follows Wilson's original development (Wilson 1971) of the renormalization-group technique, although a real-space approach was later shown to be possible also (Neimeijer and van Leeuwen 1973, 1974, Hu 1982).

For the reciprocal-space treatment, it is first necessary to transform the spin variables into reciprocal space. Applying the Fourier transformation of Eq. 5.31 to the Hamiltonian of Eq. 6.1 gives

$$\mathcal{H} = -N^{-1} \sum_{\mathbf{n}} \sum_{\mathbf{j}}' J_{\mathbf{j}} \int d\mathbf{q}\, S_{\mathbf{q}} \exp(-i\mathbf{q} \cdot \mathbf{n}) \int d\mathbf{q}'\, S_{\mathbf{q}'} \exp(-i\mathbf{q}' \cdot (\mathbf{n}+\mathbf{j}))$$

$$= -N^{-1} \int d\mathbf{q}\, S_{\mathbf{q}} \int d\mathbf{q}'\, S_{\mathbf{q}'} \left(\sum_{\mathbf{n}} \exp[-i(\mathbf{q}+\mathbf{q}') \cdot \mathbf{n}] \right)$$

$$\times \left(\sum_{\mathbf{j}}' J_{\mathbf{j}} \exp(-i\mathbf{q}' \cdot \mathbf{j}) \right)$$

$$= -\tfrac{1}{2} \int d\mathbf{q}\, S_{\mathbf{q}} S_{-\mathbf{q}} J(\mathbf{q}) \tag{6.5}$$

since the sum over $\mathbf{n}$ gives rise to a delta function in $(\mathbf{q}+\mathbf{q}')$. $J(\mathbf{q})$ is defined by the equation

$$J(\mathbf{q}) = \sum_{\mathbf{j}} J_{\mathbf{j}} \exp(i\mathbf{q} \cdot \mathbf{j}) \tag{6.6}$$

The factor of $\tfrac{1}{2}$ comes from the restriction on the initial sum that each pair of interacting spins is to be included only once.

For the Gaussian weighting term,

$$\sum_{\mathbf{n}} S_{\mathbf{n}}^2 = N^{-1} \sum_{\mathbf{n}} \iint d\mathbf{q}\, d\mathbf{q}'\, S_{\mathbf{q}} S_{\mathbf{q}'} \exp[i(\mathbf{q}+\mathbf{q}') \cdot \mathbf{n}]$$

$$= \int d\mathbf{q}\, S_{\mathbf{q}} S_{-\mathbf{q}} \tag{6.7}$$

The integrals in Eqs. 6.5 and 6.7 are over the first Brillouin zone. Combining Eqs. 6.2, 6.3, 6.4, 6.5, and 6.7 gives

$$Z = \prod_{\mathbf{n}} \int_{-\infty}^{\infty} dS_{\mathbf{n}} \exp(\tilde{\mathcal{H}}) \tag{6.8}$$

with

$$\tilde{\mathcal{H}} = -\tfrac{1}{2} \int d\mathbf{q}\, S_{\mathbf{q}} S_{-\mathbf{q}} (-\beta J(\mathbf{q}) + b) \tag{6.9}$$

We are going to apply the scaling idea of transforming the cell size in real space. An increase of cell size by a factor l in real space will correspond to a decrease of cell size in reciprocal space by a factor of l. After several transformations, we will be concerned only with values of q that are small compared with reciprocal lattice vectors, just as in real space we dealt with distances large compared with the lattice spacing. At small values of q, $J(\mathbf{q})$ can be expanded to give

$$J(\mathbf{q}) = \sum_{\mathbf{j}} J_{\mathbf{j}}(1 - i\mathbf{q} \cdot \mathbf{j} - \tfrac{1}{2}(\mathbf{q} \cdot \mathbf{j})^2 + \cdots \tag{6.10}$$

Let us assume interactions $J_{\mathbf{j}}$ confined to nearest neighbors that are located at $\pm a$ along each of the (orthogonal) axes in d dimensions. If this nearest-neighbor interaction has strength J, then the center of symmetry removes the imaginary term in $J(\mathbf{q})$ and

$$J(\mathbf{q}) = (2d - q^2 a^2)J \qquad (6.11)$$

A particularly simple lattice has been chosen with just nearest-neighbor interactions, but if universality is correct this choice will not affect the critical properties so long as the interactions are of short range. For long-range interactions, the expansion in powers of q of $J(\mathbf{q})$ (Eq. 6.10) will no longer be good, because the sums involved will not converge. This is how long-range interactions change the character of the solution for critical properties.

Substituting Eq. 6.11 into Eq. 6.9 gives

$$\bar{\mathcal{H}} = -\tfrac{1}{2}\int d\mathbf{q}\, S_\mathbf{q} S_{-\mathbf{q}}(\beta J a^2 q^2 - 2d\beta J + b)$$

$$= -\tfrac{1}{2}\int d\mathbf{q}\, \sigma_\mathbf{q}\sigma_{-\mathbf{q}}(a^2 q^2 + r) \qquad (6.12)$$

where

$$\sigma_\mathbf{q} = (\beta J)^{1/2} S_\mathbf{q} \qquad (6.13)$$

and

$$r = b/(\beta J) - 2d \qquad (6.14)$$

Our basic assumption in applying a transformation of scale is that the long-distance properties of the system dominate the behavior of the partition function and of all physical properties of interest in the critical region. This is physically plausible from the actual behavior of systems near continuous phase transitions. To apply this idea, suppose that the block length is increased by a factor l in real space; then in reciprocal space there is a corresponding length reduction by a factor l. The integral over the ith component of $\mathbf{q}$ needs to be evaluated between limits of zero and $\pi(aL)^{-1}$, and is now to be replaced by two integrals

$$\int_0^{\pi(aL)^{-1}} dq_i \to \int_0^{\pi(aL)^{-1}l^{-1}} dq_i + \int_{\pi(aL)^{-1}l^{-1}}^{\pi(aL)^{-1}} dq_i$$

If the small values of q_i dominate the physical properties, then only the first of the pair of new integrals needs to be retained and

$$\bar{\mathcal{H}} = -\tfrac{1}{2}\prod_i \int_0^{\pi(aL)^{-1}l^{-1}} dq_i\, \sigma_\mathbf{q}\sigma_{-\mathbf{q}}(a^2 q^2 + r)$$

This gives the partition function when the scale is changed by a factor of l. The renormalization group looks for a scale change of the parameters of $\bar{\mathcal{H}}$ to make it look the same as the original $\bar{\mathcal{H}}$. Two such changes of parameters are performed, one to give q the same range as the original q and the other to change the magnitude of the spin variable

$\sigma_{\mathbf{q}}$. These changes are

$$\tilde{\mathbf{q}} = l\mathbf{q} \quad \text{and} \quad \sigma_{\mathbf{q}} = \rho\tilde{\sigma}_{\tilde{\mathbf{q}}} = \rho\tilde{\sigma}_{l\mathbf{q}} \tag{6.15}$$

In terms of these variables, since $\mathbf{q}$ has dimensionality d,

$$\tilde{\mathcal{H}} = -\tfrac{1}{2}\rho^2 l^{-d} l^{-2} \int d\tilde{\mathbf{q}} \ \tilde{\sigma}_{\tilde{\mathbf{q}}} \tilde{\sigma}_{-\tilde{\mathbf{q}}} (\tilde{q}^2 a^2 + l^2 r)$$

The transformation leave $\tilde{\mathcal{H}}$ unchanged if

$$\rho^2 l^{-d} l^{-2} = 1$$

and if

$$\tilde{r} = l^2 r \tag{6.16}$$

It follows that, at the critical point,

$$\rho_c = l^{1+d/2} \quad \text{and} \quad r_c = 0 \tag{6.17}$$

Substituting this into Eq. 6.14 at the critical point gives

$$0 = \frac{bk_B T_c}{J} - 2d \quad T_c = \frac{2dJ}{bk_B} \tag{6.18}$$

At temperatures other than the critical temperature a renormalization-group transformation can still be performed, but there will be no fixed point. The transformation has two parameters, r and ρ, and the latter of these is temperature independent. Substitution of Eq. 6.18 into Eq. 6.14 shows that

$$r = \frac{2dT}{T_c} - 2d$$

$$= 2dt \tag{6.19}$$

r is directly proportional to t, and substitution in to Eq. 6.16 gives

$$\tilde{t} = l^2 t$$

In Chapter 5 the scaling parameter y was defined by the equation

$$\tilde{t} = l^y t$$

Thus

$$y = 2 \quad \text{and} \quad \nu = y^{-1} = \tfrac{1}{2} \tag{6.20}$$

This is the same as is predicted by the Ginzburg–Landau theory. We will generate one more result for the Gaussian model by recalling Eq. 5.32 for the correlation function at the critical point:

$$\hat{C}(\mathbf{q}, 0, 0) = \int d\mathbf{q}' \langle \mathbf{S}_{\mathbf{q}'} \cdot \mathbf{S}_{\mathbf{q}} \rangle$$

This is similar to the form of the equation for $\tilde{\mathcal{H}}$ in Eq. 6.12 with the term $(a^2 q^2 + r)$ left out. The same renormalization group transformation can be done, to yield

$$\hat{C}(\mathbf{q}, 0, 0) = \rho^2 l^{-d} C(l\mathbf{q}, 0, 0)$$

and substitution of $l\,|\mathbf{q}| = 1$ and $\rho = l^{1+d/2}$ (Eq. 6.17) yields

$$\hat{C}(\mathbf{q}, 0, 0) = \mathbf{q}^{-2} C(\mathbf{1}, 0, 0)$$

From the definition of η it is predicted (Eq. 5.34) that

$$\hat{C}(\mathbf{q}, 0, 0) \sim |\mathbf{q}|^{\eta-2}$$

So that

$$\eta = 0 \qquad (6.21)$$

Now that two critical exponents have been derived, the others can be calculated via the scaling laws. The arguments based in this section involve the assumptions used in the derivation of the scaling laws (Chapter 5), so the scaling laws must hold. The critical exponents that have been derived ($\nu = 0.5$, $\eta = 0$) are independent of the dimensionality and in fact correspond to the Ginzburg–Landau model (cf. Table 5.1).

6.2. Beyond the Gaussian Model

The Gaussian model is too crude and we must go beyond it in the application of the renormalization group in order to produce new and interesting results.

The technique for improvement is to advance beyond the Gaussian weighting function to the more realistic function of Eq. 6.4. Thus, we put

$$\tilde{\mathcal{H}} = \tilde{\mathcal{H}}_G + \tilde{\mathcal{H}}_I \qquad (6.22)$$

where $\tilde{\mathcal{H}}_G$ corresponds to the Gaussian model and is given by Eq. 6.12. $\tilde{\mathcal{H}}_I$ is given by

$$\tilde{\mathcal{H}}_I = -u \sum_{\mathbf{n}} S_{\mathbf{n}}^4$$

which corresponds to the fourth-power term in the weighting function. Fourier transformation gives

$$\tilde{\mathcal{H}}_I = \iiint\int d\mathbf{q}_1\, d\mathbf{q}_2\, d\mathbf{q}_3\, d\mathbf{q}_4\, S_{\mathbf{q}_1} S_{\mathbf{q}_2} S_{\mathbf{q}_3} S_{\mathbf{q}_4}\, \delta(\mathbf{q}_1 + \mathbf{q}_2 + \mathbf{q}_3 + \mathbf{q}_4)$$

A solution is now sought to the renormalization group equations with $\tilde{\mathcal{H}}_I$ treated as a perturbation of $\tilde{\mathcal{H}}_G$. The solution goes beyond the level of presentation of this book, though it uses the same ideas as were used to solve the Gaussian model; readers wishing to see a full treatment are refered to the article by Wilson and Kogut (1974). This shows that, for $l = 2$, the transformed Hamiltonian has the same form as the initial

Hamiltonian if

$$\tilde{r} = 4[r + 3cu/(1+r)] + O(u^2)$$

and (6.23)

$$\tilde{u} = 2^{4-d}[u - 9cu^2(1+r)^{-2}] + O(u^3)$$

with c not determined, except that it must be greater than zero.

Since the approach uses perturbation theory, it is necessary that u and $\tilde{u}$ always be small (compared with unity) for the theory to be valid. Furthermore, u must not be less than zero for the weighting function (Eq. 6.4) to be physically reasonable and not diverge at large S. This leads us to note that the term in the square brackets in the second of the equations above must be less than u, so that if

$$2^{4-d} \leq 1 \qquad (6.24)$$

then

$$\tilde{u} < u \quad \text{with } u \text{ and } \tilde{u} > 0$$

so that each transformation reduces the value of u and many transformations will serve to reduce u to zero. This then gives the Gaussian model. The condition 6.24 is equivalent to $d \geq 4$, so we can conclude that the Gaussian model will hold in four or more dimensions. Unfortunately, this result cannot be checked experimentally.

In four or more dimensions, we started on the smallest scale with a partition function that included the variable u, but after repeated applications of the renormalization group this variable becomes unimportant and the partition function reverts to the partition function for a Gaussian model. This leads to the idea of *relevant* and *irrelevant variables*. An irrelevant variable tends to zero after repeated transformations, while a relevant variable does not. Such behavior indicates how universality might come about: only the variables specified by universality are relevant and all other are irrelevant.

We saw in the last chapter that hyperscaling seems to break down for more than four dimensions and now the renormalization group is showing that the character of the solution changes when there are more than four dimensions. The solution reverts to the Gaussian model, which gives critical exponents v and η that follow the Ginzburg–Landau model. We conclude that the solution of the Ising model (which is where we started) in more than four dimensions gives the Ginzburg–Landau model and that hyperscaling fails. In fewer than four dimensions, hyperscaling seems to hold and the Ginzburg–Landau model does not. In exactly four dimensions, both hyperscaling and the Ginzburg–Landau theory are good; this is known as the *upper borderline dimensionality*. The *lower borderline dimensionality* is the dimensionality below which there is no phase transition to an ordered state; for $D = 1$ this dimensionality is 2.

An interesting feature of the equations that have been generated (Eq. 6.23) is that the real-space dimensionality is not confined to real-world situations of $d = 1$, 2, or 3, or even to situations in which the dimensionality is integral. There is nothing in the mathematics to prevent us treating d as a continuous variable! Indeed, such a treatment is a tempting proposition, since the solution is known in four dimensions and one could expand about this dimensionality using the parameter

$$\epsilon = 4 - d \tag{6.25}$$

This is known as the ϵ *expansion* (Wilson and Fisher 1972).

Let us now look for fixed points of Eq. 6.23 when u and r are nonzero; that is, we look for solutions other than those given by the Gaussian model. Such a fixed point corresponds to

$$\tilde{r} = r = r_c \neq 0$$

and $\tag{6.26}$

$$\tilde{u} = u = u_c \neq 0$$

Note that what was originally a perturbation, u, of arbitrary, though small, size becomes a perturbation of fixed size, u_c, at the critical point. Substituting Eq. 6.26 into Eq. 6.23 and dropping higher-order terms in u gives

$$r_c = -4cu_c(1 + r_c)^{-1} \tag{6.27}$$

and

$$1 - 2^{-\epsilon} = 9cu_c(1 + r_c)^{-2} \tag{6.28}$$

We can solve this numerically for three dimensions to get $r_c = -0.182$ and $cu_c = 0.037$, and for two dimensions to get $r_c = -0.250$ and $cu_c = 0.047$. However, the location of the fixed points is relatively uninteresting; what we want to find is the critical exponents, and these are determined by the behavior near to fixed points rather than by the location of the fixed points.

This line will be pursued in the framework of the ϵ expansion, where we assume both r_c and cu_c to be small and determine the leading term, which is of order ϵ. If r_c is small, $(1 + r_c)$ can be replaced by 1 and Eq. 6.27 becomes

$$2^{-\epsilon} = 1 - 9cu_c$$

$$-\epsilon \ln 2 = \ln(1 - 9cu_c)$$

$$= -9cu_c + O((cu_c)^2)$$

$$cu_c = \frac{\epsilon(\ln 2)}{9}$$

and

$$r_c = -4(\ln 2)\frac{\epsilon}{9} + O(\epsilon^2)$$

$$= -0.31\epsilon + O(\epsilon^2)$$

The Renormalization Group

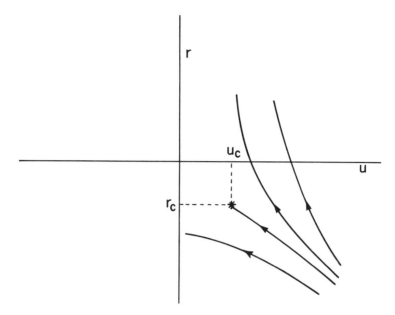

Figure 6.1. Plot of the locus of the variables u and r when renormalization group transformations are applied. If the initial values (u_0, r_0) are on a certain line, corresponding to $T = T_c$, there will be convergence to the fixed point (u_c, r_c); otherwise there is no convergence to a fixed point.

We now look at the nature of the convergence to the fixed point (u_c, r_c) when we start from some initial point (u_0, r_0). Only if the point (u_0, r_0) is on a certin line will the series converge to the fixed point (u_c, r_c); otherwise the series diverges, as illustrated in Figure 6.1. This behavior arises because T_c is a fixed parameter and there can only be convergence to the fixed point if $T = T_c$. For any initial value of u_0, r_0 must have a certain fixed value since the temperature is fixed.

Suppose that the temperature is near to T_c and l iterations have been performed to give values r_l and u_l. We do one more iteration, and if we are close to the fixed point

$$r_l \sim r_{l+1} \sim r_c$$

$$u_l \sim u_{l+1} \sim u_c$$

As is usual near a critical point, we transform to the variables $r_l - r_c$ and $u_l - u_c$. Then

$$r_{l+1} = 4[r_l + 3cu_l(1 + r_l)^{-1}]$$

and

$$r_c = 4[r_c + 3cu_c(1 + r_c)^{-1}]$$

so

$$r_{l+1} - r_c = 4\{(r_l - r_c) + 3c(u_l - u_c)(1 + r_c)^{-1} + 3cu_l[(1 + r_c)^{-1} - (1 + r_l)^{-1}]\}$$

Now

$$u_l[(1+r_c)^{-1} - (1+r_l)^{-1}] = u_l(r_c - r_l)(1+r_c)^{-1}(1+r_l)^{-1}$$
$$\sim u_c(r_c - r_l)(1+r_c)^{-2}$$

This approximation is good if $|r_c| \ll 1$, which is the case if ϵ is small. Then

$$r_{l+1} - r_c = [4 - 12cu_c(1+r_c)^{-2}](r_l - r_c) + 12c(1+r_c)^{-1}(u_l - u_c)$$
$$= M_{11}(r_l - r_c) + M_{12}(u_l - u_c)$$

In a similar way

$$u_{l+1} - u_c = 2^\epsilon 18cu_c^2(1+r_c)^{-3}(r_l - r_c)$$
$$+ 2^\epsilon[1 - 18cu_c(1+r_c)^{-2}](u_l - u_c)$$
$$= M_{21}(r_l - r_c) + M_{22}(u_l - u_c)$$

This is a matrix equation:

$$\begin{bmatrix} r_{l+1} - r_c \\ u_{l+1} - u_c \end{bmatrix} = \begin{bmatrix} M_{11} & M_{12} \\ M_{21} & M_{22} \end{bmatrix} \begin{bmatrix} r_l - r_c \\ u_l - u_c \end{bmatrix}$$

If n iterations are performed,

$$\begin{bmatrix} r_{l+n} - r_c \\ u_{l+n} - u_c \end{bmatrix} = (M)^n \begin{bmatrix} r_l - r_c \\ u_l - u_c \end{bmatrix}$$

and after many iterations the result will be dominated by the largest eigenvalue of M. Now look at the matrix M for small ϵ,

$$M_{11} \sim 4 + O(\epsilon)$$
$$M_{12} \sim 12c$$
$$M_{21} \sim 2\epsilon^2(9c)^{-1}$$
$$M_{22} \sim 1 + O(\epsilon)$$

The matrix has eigenvalues M_{11} and M_{22} and the largest eigenvalue is M_{11}. After a large number of iterations, l,

$$r_{l+1} - r_c = M_{11}(r_l - r_c) = (4 - 12cu_c(1+r_c)^{-2})(r_l - r_c)$$

We are now in a position to calculate the critical exponent v using the same line of reasoning as in Section 6.2: t is replaced by $(r - r_c)$ and, using the same steps in the argument, we get

$$\bar{t} = M_{11} t$$
$$\bar{t} = 2^y t$$

Since

$$y = v^{-1} \qquad v = \ln 2 (\ln M_{11})^{-1}$$

Now,
$$M_{11} = 4 - 12cu_c(1+r_c)^{-2}$$
$$= 4 - 4\epsilon(\ln 2)/3 + O(\epsilon^2)$$
$$\ln(M_{11}) = \ln 4 - \epsilon(\ln 2)/3 + O(\epsilon^2)$$
$$\nu = \frac{1}{2 - \epsilon/3}$$
$$= \frac{1}{2} + \frac{\epsilon}{12} + O(\epsilon^2)$$

It is physically appropriate that the term $\ln 2$ has canceled out as the two came from the length-scaling factor, which should be irrelevant. The result is also independent of the parameter c.

For the three-dimensional Ising model, the critical exponent ν is 0.58, while the best estimates give a value of 0.63 (Table 5.1). The next term in the ϵ expansion has in fact been calculated, and in three dimensions gives $\nu = 0.604$. The expansion should not be expected to give good results with ϵ as large as unity, but the result is clearly better than the Ginzburg–Landau theory.

In fact, the calculation can be done without having to approximate to small ϵ and for spin dimensionalities other than 1 (Le Guillou and Zinn-Justin 1980, Baker et al. 1978, and references therein). The results of such calculations were given in Table 5.1.

Suggested Further Reading

Wilson and Kogut (1974)
Hu (1982)
Fisher (1982)
Pfeuty and Toulouse (1977)

7

CRITICAL DYNAMICS

We have already seen that there is a slowing down of dynamic processes near critical points, and that response times tend to infinity at the critical point. These phenomena can alternatively be described in terms of certain transport properties becoming infinite at the critical point. It is found experimentally that these divergences follow simple power laws, just as in the case of static properties, so that critical exponents may be defined for dynamic properties. These exponents only follow universality if universality classes are expanded to specify conservation laws as well as the conditions for universality of static exponents.

In this chapter we show that the scaling laws can be extended to include dynamic properties (Ferrell et al. 1968, Halperin and Hohenberg 1969a). First we describe some special features of the dynamics of the Ising model where the z component of spin on any atom, i, commutes with the Hamiltonian

$$(\mathcal{H}_{\text{ISING}}, S_i^z) = 0$$

This implies that, if any critical configuration of the z component of the spins is set up, that configuration is conserved and there are no dynamics at all. In fact, there is no way for a pure Ising system to come into thermal equilibrium! This is clearly not realistic, and some small perturbation terms will have to be added to the Hamiltonian just to ensure that it comes to equilibrium. It will then be necessary to consider how the dynamics are influenced by the nature of the perturbation.

This difficulty applies only to the Ising model, and our other standard models can attain equilibrium without recourse to any perturbations. However, the dynamic problem is harder to solve than the static problem and we find that there are no exact solutions, even in one dimension ($d = 1$). All this implies that we will be building on less-solid foundations than has been the case in previous chapters.

7.1. Dynamic Scaling

In setting up a framework for the discussion of dynamical problems, we need to discuss the time or frequency dependence of the critical fluctuations. This is achieved by generalizing the correlation function formalism introduced in Chapter 5 to include time-dependent correlations. If we introduce the variable τ for time (so as to distinguish time from reduced temperature t), then Eq. 5.30 may be generalized to an equation for a time-dependent correlation function. If we put $h = 0$, we get

$$C(\mathbf{q}, t, \tau) = \sum_{\mathbf{R}} e^{i\mathbf{q}\cdot\mathbf{R}}[\langle \mathbf{S_0}(0) \cdot \mathbf{S_R}(\tau)\rangle - \langle \mathbf{S_0}\rangle\langle \mathbf{S_R}\rangle] \quad (7.1)$$

where we assume that the average value of $\langle \mathbf{S_R}\rangle$ is time independent and where $\mathbf{S_R}(\tau)$ is the value of $\mathbf{S}$ on site $\mathbf{R}$ at time τ. Fourier transformation of this function so as to make frequency rather than time a variable gives

$$\hat{C}(\mathbf{q}, t, \omega) = (2\pi)^{-1}\int_{-\infty}^{\infty} \exp(i\omega\tau)\,\hat{C}(\mathbf{q}, t, \tau)\,d\tau \quad (7.2)$$

It is usual to split this function into a static part $\hat{C}(\mathbf{q}, t)$, as given by Eq. 5.30, and a dynamic part. So that

$$\hat{C}(\mathbf{q}, t, \omega) = \hat{C}(\mathbf{q}, t)\,F(\mathbf{q}, t, \omega)\,d(\omega) \quad (7.3)$$

The dynamic part consists of a spectral weight function $F(\mathbf{q}, t, \omega)$ and a detailed balance term, $d(\omega)$ given by (Lovesey 1984)

$$d(\omega) = \frac{\omega}{1 - \exp(-\hbar\omega\beta)}$$

In the critical region, $|\hbar\omega\beta| \ll 1$ so that $d(\omega)$ is a constant, $(\hbar\beta)^{-1}$. The inclusion of the detailed balance term ensures that the spectral weight function is an even function of ω at all $\mathbf{q}$ and t.

The spectral weight function is normalized such that

$$\int_{-\infty}^{\infty} F(\mathbf{q}, t, \omega)\,d\omega = 1 \quad (7.4)$$

It expresses the frequency dependence of the correlation with wavevector $\mathbf{q}$ at temperature t.

In the same spirit as in scaling and renormalization group theory, we postulate that each block in reciprocal space of size $\mathbf{q}$ at temperature t in zero magnetic field has a characteristic frequency $\omega_c(\mathbf{q}, t)$. A convenient definition of the characteristic frequency is the median frequency, which divides the area under the spectral weight function in half; that is, we define ω_c by

$$\int_{-\omega_c}^{\omega_c} F(\mathbf{q}, t, \omega)\,d\omega = \tfrac{1}{2} \quad (7.5)$$

In scaling theory, we work in terms of dimensionless variables, so on the right-hand side of Eq. 7.3 we will replace ω by the dimensionless quantity ω/ω_c ($\mathbf{q}$ and t are already dimensionless) to get (remembering the normalization Eq. 7.4)

$$\hat{C}(\mathbf{q}, t, \omega) = \hbar\beta\omega_c^{-1}\hat{C}(\mathbf{q}, t)\,F(\mathbf{q}, t, \omega/\omega_c) \quad (7.6)$$

In static scaling, t^ν scales as q (cf. Eq. 5.33) and we postulate a similar relationship in dynamic scaling, so that

$$F(\mathbf{q}, t, \omega/\omega_c) = F(\mathbf{q}t^{-\nu}, \omega/\omega_c)$$

This can be checked experimentally, as we will see later.

In dynamic scaling theory, we make a similar sort of postulate for the scaling properties of the characteristic frequency, ω_c, as we did in static scaling for the scaling properties of the Gibbs free energy and of the correlation length. In fact, we assume that

$$\omega_c(\mathbf{q}, t) = q^z \omega_c(\tilde{\mathbf{q}}, \tilde{t})$$

Now we assume that t^ν scales as q for $\omega_c(\mathbf{q}, t)$, just as we assumed it did for the spectral weight function. Then

$$\omega_c(\mathbf{q}, t) = q^z f(\mathbf{q}t^{-\nu}) \tag{7.7}$$

Because the characteristic frequency depends on the conservation laws as well as on the real-space and spin dimensionalities, z will also depend on the conservation laws. This is why the extra condition has to be added to universality when dynamic properties are under consideration.

We note that at the critical point ($t = 0$) the characteristic frequency is given by

$$\omega_c(\mathbf{q}, 0) = q^z f(0) \qquad (T = T_c) \tag{7.8}$$

The predictions of Eqs. 7.7 and 7.8 can be verified experimentally. We consider two cases in detail.

7.2. Heisenberg Ferromagnet

At small wavevectors below the critical temperature, we know that the excitation spectrum is comprised of magnons, and the characteristic frequency will be the magnon frequency given by

$$\hbar\omega_c(\mathbf{q}, t) = D(t)q^2 \tag{7.9}$$

where $D(t)$ is called the *spin-wave stiffness constant*. There is a subtlety to be considered here, however, which is that the magnons are excitations in the x–y plane, if z is the magnetization direction, and Eq. 7.9 can only apply to correlations in this plane. Our initial definition of the spin correlation function in Eq. 5.22 assumed isotropy in spin space and this needs to be generalized by writing

$$C^{\alpha\beta}(\mathbf{R}, t, h) = \langle (S_\mathbf{0}^\alpha - \langle S_\mathbf{0}^\alpha \rangle)(S_\mathbf{R}^\beta - \langle S_\mathbf{R}^\beta \rangle) \rangle \tag{7.10}$$

where α and β equal x, y, or z for the Heisenberg model, which uses spin vectors in three dimensions. We must now put this into the rest of the scaling derivation by giving the functions $\hat{C}(\mathbf{q}, t)$, $\hat{f}(\mathbf{q})$, $F(\mathbf{q}, t, \omega)$, $\hat{C}(\mathbf{q}, t, \omega)$, and $\omega_c(\mathbf{q}, t)$ the superscripts α and β, but in fact all the same equations will apply in generalized form after this has been done. We have chosen not to use these superscripts except where they are needed, so as to keep the notation simple. In fact, they are not needed at and above the critical temperature in zero magnetic field, because the system is isotropic in spin space and it is only below the critical temperature that their introduction is necessary. Equations 7.7 and 7.9 combine to give

$$\hbar q^z f(\mathbf{q}t^{-\nu}) = D(t)q^2$$

The terms in q only balance if

$$f(\mathbf{q}t^{-\nu}) \sim (\mathbf{q}t^{-\nu})^{2-z}$$

In this case

$$D(t) \sim t^{(z-2)\nu} \qquad (T < T_c) \tag{7.11}$$

The spin-wave stiffness constant varies with temperature as $t^{(z-2)\nu}$. The temperature variation of this constant is predicted independently of dynamic scaling by two theories: hydrodynamic theory (Halperin and Hohenberg 1969b), which is a macroscopic approach; the mode–mode coupling theory (Fixman 1962, Kadanoff and Swift 1968, Kawasaki 1967, 1970, 1976), which is a microscopic approach. These predict a critical slowing down of the spin wave energies such that

$$D(t) \sim t^{\nu - \beta} \qquad (T < T_c) \tag{7.12}$$

Comparison with Eq. 7.11 shows that

$$(z - 2)\nu = \nu - \beta, \qquad z = 3 - \beta/\nu \tag{7.13}$$

Reference to the values of β and ν in Table 5.1 shows that for the Ginzburg–Landau model $z = 2.5$ and for the Heisenberg ferromagnet in three dimensions $z = 2.486$.

An alternative form combining Eq. 7.13 with the hyperscaling laws (Eqs. 5.16 and 5.28) leads to the expression

$$z = 4 - \frac{d}{2} - \frac{\eta}{2} \tag{7.14}$$

Above T_c in the Heisenberg ferromagnet, the relevant characteristic frequency will be Λq^2, where Λ is the spin-diffusion constant (Halperin and Hohenberg 1969b, Kawasaki 1976), so that

$$\hbar \omega_c(\mathbf{q}, t) = \Lambda(t)q^2 \tag{7.15}$$

This has the same form as Eq. 7.9, so that $\Lambda(t)$ will have the same form as $D(t)$, owing to the symmetry of the scaling laws above and below T_c (cf. Eq. 5.13), and there will be a critical slowing down with

$$\Lambda(t) \sim (-t)^{\nu - \beta} \qquad (T > T_c) \tag{7.16}$$

7.3. Heisenberg Antiferromagnet

At small wavevectors below the critical temperature, there are spin wave excitations in the transverse plane with frequency given by

$$\hbar \omega_c(\mathbf{q}, t) = c(t)q \tag{7.17}$$

Combination of Eqs. 7.17 and 7.7 gives

$$\hbar q^z f(\mathbf{q}t^{-\nu}) = c(t)q$$

The terms in q only balance if

$$f(\mathbf{q}t^{-\nu}) \sim (qt^{-\nu})^{1-z} \tag{7.18}$$

In this case

$$c(t) \sim t^{(z-1)\nu} \qquad (T < T_c) \tag{7.19}$$

Hydrodynamic theory (Halperin and Hohenberg 1969b) predicts a critical slowing down of the spin-wave energy such that

$$c(t) \sim t^{\nu/2} \qquad (T < T_c) \tag{7.20}$$

Comparison between Eqs. 7.19 and 7.20 shows that

$$z = 1.5 \tag{7.21}$$

Above T_c the antiferromagnetic reciprocal lattice points cease to have physical existence, as the lattice symmetry changes so as to make all sites equivalent. This means that there is no reason for the characteristic width to go to zero at this point. In fact, at small wavevectors in the antiferromagnetic zone, the characteristic frequency will be a constant, independent of q, since the antiferromagnetic zone center corresponds to a zone boundary in the paramagnetic lattice.

$$\omega_c(\mathbf{q}, t) = \Gamma_0(t) \tag{7.22}$$

Comparison with Eq. 7.7 for $z = 1.5$ shows that

$$\Gamma_0(t) = q^{1.5} f(\mathbf{q}t^{-\nu})$$

The dependence on q of the right-hand side of this equation disappears if

$$f(\mathbf{q}t^{-\nu}) \sim (qt^{-\nu})^{-1.5}$$

so that

$$\Gamma_0(t) \sim t^{1.5\nu} \tag{7.23}$$

This gives a critical slowing down with exponent 1.5ν, or 1.06 if we put $\nu = 0.705$ for the Heisenberg model (cf. Table 5.1).

Dynamic scaling makes many predictions that can be checked experimentally. We will see later that the theory agrees very well with experimental measurements of dynamic correlation functions and of characteristic frequencies in the critical region by neutron scattering. As with static scaling, however, we have to note that dynamic scaling does not solve any dynamical problem absolutely; it merely relates the solution at point $(\mathbf{q}, t, \omega)$ to that at another point in terms of the three scaling exponents, x, y, and z. Recourse to hydrodynamic theory or to mode–mode coupling theory allows the exponent z to be determined for many universality groups (Riedel and Wegner 1970, Finger 1977), as well as some limiting forms for ω_c.

Approximate theories have been formulated (Résibois and Piette 1970, Hubbard 1971) that predict the form of the spectral weight function and these compare well with experimental data. It has also proved possible to apply renormalization-group techniques to dynamic problems (Halperin and Hohenberg 1977, articles by J. D. Gunton, M. Suzuki and G. F. Mazendko in Enz 1979, Folk and Iro, 1985).

Suggested Further Reading

Halperin and Hohenberg (1969a, 1977)
Stanley (1971)
Forster (1975)

8

MORE COMPLEX MAGNETIC SYSTEMS

8.1. Crossover Exponents

Most real systems do not have Hamiltonians that are exactly those set out for the model Hamiltonians in Chapter 4. A common situation is for the large term in the Hamiltonian to correspond to the Heisenberg model (Eq. 4.3) but for there to be another, smaller, term that tends to align the spins in one particular direction (a *uniaxial* term) or to confine the spins to one particular plane (an *easy-plane* term). These *anisotropic* effects can arise from crystal-field, dipolar, or anisotropic exchange interactions; they lead to an effective lowering of the spin dimensionality, D, since they tend to confine the order parameter (the magnetization) to a lower number of dimensions that does the predominant Heisenberg term.

Let us suppose the Heisenberg Hamiltonian to be $\mathcal{H}_1(J)$, where J is the exchange parameter. We assume for simplicity just a single nearest-neighbor exchange interaction. Now let the small term corresponding to an Ising or an $X-Y$ Hamiltonian be $\mathcal{H}_2(gJ) = g\mathcal{H}_2(J)$, where $g \ll 1$. There will be two characteristic energies of the system:

$$E_1 = zJS(S+1) \simeq k_B T_c$$

and

$$E_2 = gzJS(S+1) \simeq k_B T_A$$

where z is the number of nearest neighbors. At very low temperatures ($T \ll T_A$) the magnetization will be confined to one or two dimensions because there is not the thermal energy to overcome the anisotropy energy E_2. At higher temperatures ($T \gg T_A$) the anisotropy will become only a small perturbation and the system will behave according to the Heisenberg Hamiltonian. There is a crossover in effective spin dimensionality at around a temperature of T_A.

A similar crossover happens in the critical region. Very close to the critical point where $|T - T_c|$ is small compared with T_A, the critical fluctuations will be confined to one or two dimensions. Farther away from T_c, however, where $|T - T_c|$ is large compared with T_A, there will be enough energy available for the critical fluctuations to have a magnetization that varies in three dimensions. The critical properties cross over from one universality class to another at a temperature of the order of $T_c \pm T_A$. Figure 8.1 illustrates this idea with a crossover from Heisenberg to Ising behavior in the susceptibility.

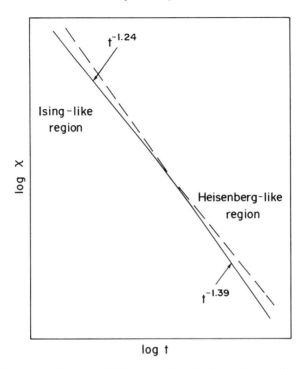

Figure 8.1. Crossover of the susceptibility from Ising-like behavior very close to the critical point ($\chi \sim t^{-1.24}$) to Heisenberg-like behavior further away from the critical point ($\chi \sim t^{-1.39}$).

Figure 8.2 shows the effect of the crossover on the inverse correlation length, $\kappa_1 = 1/\xi$. Very close to T_c in the Ising-like region the correlation will diverge for the z components of the spin following the expected Ising behavior with $\kappa_1^{zz} \sim t^{0.63}$ (cf. Table 5.1), while the xx and yy correlations will not diverge. Farther away from T_c in the Heisenberg-like region the fluctuations will be isotropic with $\kappa_1 \sim t^{0.71}$, as is to be expected for a Heisenberg system.

The crossover behavior is such that, closest to the critical point, the spin dimensionality has its lowest value and the real-space dimensionality has its highest value. These are features that tend to stabilize the low-temperature phase.

We now describe how crossovers can be dealt with in scaling theory (Riedel and Wagner 1969, Pfeuty et al. 1974). We treat the anisotropy parameter g as being small and describe the critical temperature as a function $T_c(g)$ of g. For the unperturbed system ($g = 0$) we define a reduced temperature as before by

$$t = \frac{T - T_c(0)}{T_c(0)} \tag{8.1}$$

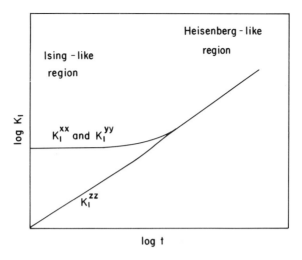

Figure 8.2. Logarithmic plot of the variation of the inverse correlation length, κ, with reduced temperature, t, for a system that crosses over from Ising to Heisenberg behavior. At the critical point, the correlation length for zz fluctuations becomes infinite, while for xx or yy fluctuations it tends to a constant value. Far from the critical point, the correlations cross over to become isotropic, as is appropriate for a Heisenberg system.

and for the perturbed system we define a reduced temperature $\dot{t}$ by

$$\dot{t} = \frac{T - T_c(g)}{T_c(0)} \tag{8.2}$$

The shift in critical temperature caused by the anisotropy is defined by

$$t_s = \frac{T_c(g) - T_c(0)}{T_c(0)} = t - \dot{t} \tag{8.3}$$

We now apply scaling to the Gibbs free energy of the system. For the unperturbed Hamiltonian, we had in zero field (Eqs. 5.2 and 5.6)

$$l^d G(t) = G(\bar{t}) = G(l^\nu t)$$

The perturbation introduces an extra parameter g so that the free energy becomes $G(t, g)$ and in the perturbed system

$$l^d G(\dot{t}, g) = G(l^\nu \dot{t}, \bar{g}) \tag{8.4}$$

Now we postulate that

$$\bar{g} = l^{\phi y} g \tag{8.5}$$

which is the same sort of postulate as is usual in scaling theory (cf. Eq. 5.3). Then putting $l^\nu \dot{t} = 1$ and recalling Eq. 5.12, we find

$$G(\dot{t}, g) = \dot{t}^{2-\dot{\alpha}_G}(1, g/\dot{t}^\phi) \tag{8.6}$$

so that $\dot{t}^\phi$ scales as g. This scaling equation should be applicable in the temperature range where the critical properties are governed by the symmetry of the perturbation, that is, for

$$|\tilde{t}/t_s| \ll 1$$

We can do a similar scaling analysis for the region after the critical exponents have crossed over to those corresponding to the unperturbed Hamiltonian, that is, for

$$|\tilde{t}/t_s| \gg 1 \qquad |\tilde{t}| \ll 1$$

We postulate that after crossover the same scaling equation holds for the Gibbs free energy

$$G(t, g) = t^{2-\alpha} G(1, g/t^\phi) \tag{8.7}$$

This postulate gives t^ϕ as scaling like g. Since both t^ϕ and $\tilde{t}^\phi$ scale as g, it follows that

$$t_s = t - \tilde{t} \sim g^{1/\phi} \tag{8.8}$$

ϕ is known as the *crossover exponent* and the shift in the critical temperature caused by a perturbation g scales as $g^{1/\phi}$. For the crossover from Heisenberg to Ising symmetry, $\phi = 1.25 \pm 0.015$ (Pfeuty et al. 1974).

8.2. Tricritical Points

Let us consider a three-dimensional Ising antiferromagnetic system in a magnetic field H along z. The phase diagram of the system is as shown in Figure 8.3. In contrast to the ferromagnetic case, the application of a

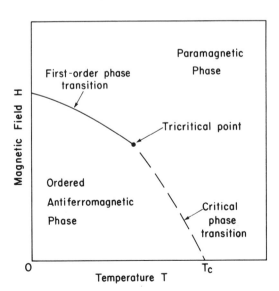

Figure 8.3. Phase diagram of an Ising antiferromagnet in a magnetic field H. There is a critical phase transition in some regions and a first-order phase transition in others. The point at which the line of first-order phase transitions changes to a line of critical phase transitions is called a *tricritical point*.

small field does not destroy the critical phase transition. The broken line in the figure traces the variation of the critical temperature with the applied field. There is a point, called the *tricritical point*, at which the phase transition changes in nature to become first order.

In the previous section, we introduced the concept of a *line of critical points* with critical temperature $T_c(g)$ that varies with some term of magnitude g in the Hamiltonian. Here we have just such a line of points with the parameter g directly proportional to the applied magnetic field H. The line changes in character at the tricritical point to become a line of first-order phase transitions. The meeting point of lines of phase transitions, when at least one of the lines involves a critical phase transition, is called a *multicritical point*. The particular multicritical point that was shown in Figure 8.3 is a tricritical point.

8.3. Ginzburg–Landau Theory of Tricritical Points

We now show how tricritical points can arise in the Ginzburg–Landau theory and we use this treatment to show that tricritical points have some special properties. The discussion follows lines set out by Aharony (1982).

In Chapter 2 we showed that the Ginzburg–Landau theory rests on the assumption that the Helmholtz free energy can be expanded about the critical point in powers of the order parameter η. We make this same expansion to sixth order (cf. Eq. 2.1), but with the expansion coefficients α now functions of g as well as of T:

$$F(t, \eta, g) = F_0(T, g) + \alpha_2(T, g)\eta^2 + \alpha_4(T, g)\eta^4 + \alpha_6(T, g)\eta^6 + \cdots \quad (8.9)$$

For equilibrium, the free energy will have a minimum value as a function of η, so that

$$\frac{\partial F}{\partial \eta} = 2\eta(\alpha_2(T, g) + 2\alpha_4(T, g)\eta^2 + 3\alpha_6(T, g)\eta^4) = 0$$

and (8.10)

$$\frac{\partial^2 F}{\partial \eta^2} = 2(\alpha_2(T, g) + 6\alpha_4(T, g)\eta^2 + 15\alpha_6(T, g)\eta^4) > 0$$

We saw in Chapter 2 that these equations lead to solutions with $\eta > 0$ for $T < T_c$ and $\eta = 0$ for $T > T_c$, and so to a critical phase transition, if α_2 changes sign at T_c and if α_4 is greater than zero. If both α_4 and α_6 are greater than zero, these same conclusions will apply when the sixth order term is included in the expansion. The introduction of the parameter g changes the situation in that T_c will be a function of g, so that Eq. 2.2 becomes

$$\alpha_2(T) = (T - T_c(g))\alpha_0 \quad (8.11)$$

with α_0 positive and there will be a line of critical points in the T–g diagram (Figure 8.3).

Near to the critical point, expansion in powers of $(T - T_c)$ yields a solution of Eq. 8.10

$$\eta^2 = \frac{\alpha_0}{2\alpha_4}(T - T_c)\left(1 + \frac{3\alpha_0\alpha_6}{\alpha_4^2}(T - T_c) + \cdots\right) \tag{8.12}$$

The order parameter still varies asymptotically as $(T - T_c)^{1/2}$ but there is a correction term (cf. Eq. 3.3) involving $\alpha_0\alpha_6(T - T_c)/\alpha_4^2$ that implies deviations from the power law away from the critical point. The magnitude of the correction term determines the range of temperatures over which the simple power law holds.

As we move along the line of critical points by varying g, we come to a tricritical point if α_4 becomes zero. Beyond this point, for negative α_4, the phase transition becomes first order. It was, of course, because of this situation of α_4 becoming zero and then negative that we needed to extend the expansion of the free energy to the sixth-power term. The sixth-order term, α_6, is assumed to remain positive, so as to ensure overall stability.

In the situation where $\alpha_4 = 0$, the solution of Eqs. 8.10 and 8.11 is

$$\eta = (\alpha_0(T - T_c)/3\alpha_6)^{1/4} \tag{8.13}$$

which corresponds to a new value of the critical exponent β of $\frac{1}{4}$. There has been a change from the behavior along the line of critical points where β is $\frac{1}{2}$ (Eq. 8.12). What happens is that as we approach the tricritical point the correction term in Eq. 8.12 becomes larger and larger ($\alpha_4 \to 0$), so that the region over which the behavior with $\beta = \frac{1}{2}$ occurs becomes narrower and narrower. Right at the tricritical point, the correction term diverges and Eq. 8.12 breaks down. In the region close to the tricritical point there will be crossover behavior from an exponent of $\frac{1}{2}$ very close to T_c to an exponent of $\frac{1}{4}$ farther away from T_c.

At the tricritical point, there is a new set of critical exponents known as *tricritical exponents*. Just as the Ginzburg–Landau predictions for critical exponents at ordinary critical points may not be appropriate, so it would not be surprising if the prediction of $\beta = \frac{1}{4}$ at the tricritical point is not quantitatively correct: however the prediction of a different exponent at a tricritical point would be expected to hold quite generally.

Now we look at how the line of first-order phase transitions occurs beyond the tricritical point. This corresponds to the region where α_4 is negative and the upper Eq. 8.10 has five solutions for $\partial F/\partial \eta = 0$:

$$\eta = 0$$

and

$$\eta^2 = \frac{-\alpha_4}{3\alpha_6}\left[1 \pm \left(1 - \frac{3\alpha_2\alpha_6}{\alpha_4^2}\right)^{1/2}\right]$$

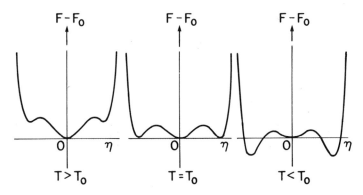

Figure 8.4. The Helmholtz free energy F plotted as a function of the order parameter η in the region where the phase transition is first order at temperature T_0. The first-order phase transition occurs when the minimum value of the free energy switches from the central minimum at $\eta = 0$ to the minima at nonzero η.

For small positive values of α_2 all these solutions are real and F has three minima and two maxima. In view of Eq. 8.11, we expect α_2 to be more sensitive to temperature than α_4 or α_6. Figure 8.4 plots the Helmholtz free energy, as given by Eq. 8.9, for three different temperatures (that is for three different values of α_2) with α_2 and α_6 positive and with α_4 negative. There are three minima in F; at high temperatures the lowest minimum is at $\eta = 0$, so the system will be disordered, while at low temperatures the lowest minima are at nonzero values of η and the system will be ordered. The center plot of Figure 8.4 corresponds to the first-order transition at temperature T_0, when all three minima correspond to the same value of the free energy. At this temperature the system goes discontinuously from the disordered to the ordered state.

At $T = T_0$ we can solve Eq. 8.9 with $F = F_0$, together with Eq. 8.10, to get the jump $\Delta\eta$ in the order parameter at the first-order phase transition as

$$(\Delta\eta)^2 = \frac{-\alpha_4}{2\alpha_6} \tag{8.14}$$

We also find the condition

$$\alpha_2 = \frac{\alpha_4^2}{4\alpha_6} \quad (T = T_0)$$

which defines the line of first-order phase transitions. At the tricritical point, $\alpha_4 = 0$, both the jump in the order parameter and α_2 go to zero as expected.

8.4. Bicritical Points

Bicritical points occur for similar situations as tricritical points when the underlying Hamiltonain is not purely Ising-like in character but when it is

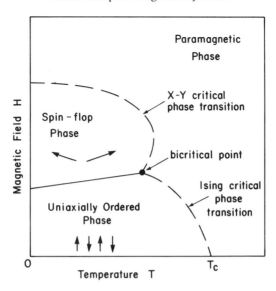

Figure 8.5. Phase diagram of a Heisenberg antiferromagnet with uniaxial anisotropy in a magnetic field H. There are two ordered phases, corresponding to uniaxial order and spin-flop order. These are separated by a line of first-order phase transitions. The point where this line meets the two lines of critical phase transitions is called a *bicritical point*.

primarily Heisenberg-like with a uniaxial anisotropy. We showed in Section 8.1 that such Hamiltonians show crossover behavior in zero applied field. When we apply a field along z, the phase diagram is as shown in Figure 8.5.

The Heisenberg term in the Hamiltonian allows for the existence of a second ordered phase known as the spin-flop phase. In this phase the spins lie antiferromagnetically in the X–Y plane; to put the spins in such a phase costs energy at the expense of the uniaxial anisotropy, but gains energy because the spins are able to cant in the direction of the field. For large enough fields, the spin-flop phase wins out over the uniaxially ordered phase. The transition between the two ordered phases is of first order.

Both ordered phases show a line of critical phase transitions to the paramagnetic phase. For the uniaxially-ordered phase this transition is from Ising-like symmetry to paramagnetism, while for the spin-flop phase it is from X–Y symmetry to paramagnetism. In fact, the two ordered phases correspond to different universality classes.

The point where the three lines of phase transitions meet is called a *bicritical point*. We would expect, by analogy with tricritical points that bicritical points exhibit critical exponents that have special values.

8.5. Lifshitz Points

We have seen that a bicritical point occurs at the triple point of a paramagnetic phase and of two ordered phases. A different type of multicritical point arises when one of the phases involves spin ordering commensurate with the lattice and the other spin-ordering that is incommensurate: the point is then called a *Liftshitz point*. In the discussion, we implicitly assumed that the ordered phases are commensurate with the lattice and have a magnetic unit cell of fixed size.

We should first introduce the idea of *incommensurate spin ordering*. For simple lattices with just nearest-neighbor interactions, the ordered state is either ferromagnetic or commensurate antiferromagnetic. When the interactions become more numerous or when the lattice becomes more complex, the nature of the ordered state becomes less obvious. Lyons and Kaplan (1960) showed that at zero temperature the ordered state need not be commensurate with the lattice: in fact, the ordered state at $T = 0$ corresponds to the wavevector $\mathbf{q}$ that gives a minimum value of the function $J(\mathbf{q})$ as defined in Eq. 6.6. For very simple cases, symmetry fixes this minimum at a particular point in the zone, but the conditions need not necessarily be very complicated for this minimum to be at arbitrary wavevector $\mathbf{q}$, so that the structure becomes incommensurate. This value of $\mathbf{q}$ will be a function $\mathbf{q}(g, T)$ of the perturbing effects g in the Hamiltonian and of the temperature. There will be a line of phase transitions corresponding to values of g and T where $\mathbf{q}$ becomes commensurate, as is shown in Figure 8.6.

The point where this line meets with the lines for the order–disorder phase transition is called a Lifshitz point. The transitions from the ordered phase to the paramagnetic disordered phase are critical in

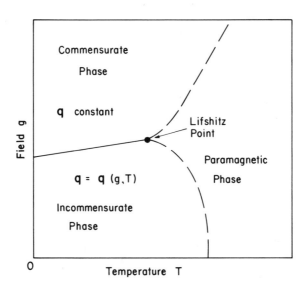

Figure 8.6. Phase diagram of a system showing a *Lifshitz point*. There are two ordered phases, one commensurate with the lattice and the other characterized by an incommensurate wavevector $\mathbf{q}(g, T)$. The Liftshitz point is the triple point of the two ordered phases and of the paramagnetic phase.

nature. It seems that the phase transition between the two ordered phases may be either first order or continuous depending on the particular system involved (Michelson 1977). As with bicritical and tricritical points, it is to be expected that critical exponents will have special values at the Lifshitz point.

Suggested Further Reading

Aharony (1982)
Fisher (1974)
Hornreich (1980)

9

DILUTION, PERCOLATION, AND RANDOM FIELDS

9.1. Dilution

All the critical systems that have been discussed so far have been homogeneous. In this chapter we look at cases in which the Hamiltonian is not the same for each magnetic atom, but instead has an intrinsic random part. A simple way to bring this about is to replace some magnetic atoms randomly by nonmagnetic atoms in the crystal lattice. This is called *dilution*; it is easy to bring about dilution experimentally, and if the diluting atom has similar size and chemical properties to the magnetic atom it is often found to be distributed randomly in the lattice.

If dilution is sufficiently great, there will be no "exchange path" right through the crystal so that there can be no long-range order. Long-range order will appear at the *percolation threshold* where there is an exchange path over an infinite distance. This will correspond to a critical concentration of magnetic atoms known as the *percolation concentration*, p_c. Figure 9.1 shows the phase diagram for a diluted magnetic system with magnetic atom concentration p. The critical temperature $T_c(p)$ goes to zero at $p = p_c$.

In this section we discuss the critical properties in the region where p is sufficiently greater than p_c that there are many exchange paths through the crystal and we are not in the "percolation region." The first question might be whether there is still a sharp critical phase transition at all in such a nonhomogeneous system. The answer, both theoretically and experimentally, seems to be that the critical phase transition remains present. Conceptually, this is justified by realizing that, close to the phase transition, there will be large regions, of size ξ, of ordering (or disordering) and that if ξ is large enough it will average out the randomness.

The next question might be whether such diluted systems have the same critical exponents as the undiluted system. The answer is that if the critical exponent α of the specific heat in the undiluted system is less than zero the critical exponents are unaffected by dilution; if α is greater than zero the critical exponents do change. This is known as the Harris criterion (Harris 1974). Of the standard models, only the Ising model in three dimension has $\alpha > 0$ (Table 5.1), so that only in that case is it predicted that dilution will change the critical exponents; γ and ν are predicted to change from 1.24 and 0.63 in the undiluted case to 1.34 and

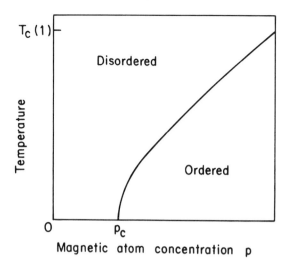

Figure 9.1. Critical temperature $T_c(p)$ of a diluted magnetic system plotted against the magnetic ion concentration, p. Above the percolation concentration p_c there is long-range order at $T = 0$.

0.68, respectively (Jug 1983). The other static critical exponents can be calculated from γ and ν via the scaling laws.

9.2. Percolation

The percolation point ($p = p_c$, $T = 0$) in Figure 9.1 is the end of a line of critical phase transitions and so represents a multicritical point. Critical exponents would be expected to take on special values at this point, as with other multicritical points.

Critical exponents at the percolation point can be defined in three ways, according to the path that is taken. First, we can take a path along the $T = 0$ axis of Figure 9.1 and denote critical exponents such as γ_p, ν_p, β_p, and η_p by the subscript p. Second, we can take a path that fixes $p = p_c$ and varies the temperature; these critical exponents such as γ_T, ν_T, and η_T are denoted by the subscript T. Third, we can approach the percolation point along the line of critical phase transitions and denote the variation of the critical temperature $T_c(p)$ with $(p - p_c)$ by the critical exponent ϕ.

The first of these sets of exponents follows from the geometry of clusters in the percolation region and is independent of the dimensionality of the order parameter or of the particular lattice type. Table 9.1 lists values for these exponents in one, two, and three dimensions (short-range interactions are assumed). The percolation concentration p_c can be calculated from the geometry of any particular lattice. Values are tabulated by Stinchcombe (1983).

Table 9.1. Percolation Critical Exponents. The Results in One Dimension are Exact; Those in Two and Three Dimensions are Taken From Appendix 1 of Essam (1980)

Dimension	γ_p	ν_p	β_p
1	1	1	0
2	2.43 ± 0.02	1.34 ± 0.02	0.139 ± 0.003
3	1.66 ± 0.07	0.83 ± 0.05	0.41 ± 0.01

Near the percolation point, the lattice consists of long chains of interacting spins with occasional nodes from which three or more chains emanate. The existence of these long chains suggest that the thermal properties will be similar to those of a one-dimensional system. From this idea, a scaling theory has been developed for the percolation region (Stauffer 1976, Stanley et al. 1976, Lubensky 1977) that predicts that the correlation length $\xi(p - p_c, T)$ is given by

$$\xi(p - p_c, T) = |p - p_c|^{-\nu_p} f(\xi_1(p - p_c)^\phi) \tag{9.1}$$

where ξ_1 is the correlation length of the one-dimensional system at temperature T.

It follows from this scaling theory that

$$\nu_p = \phi \nu_T, \qquad \gamma_p = \phi \gamma_T \qquad \eta_p = \eta_T \tag{9.2}$$

For the Ising model in any number of dimensions, $\phi = 1$ (Wallace and Young 1978) as it does also for the Heisenberg model in one dimension (Thorpe 1975). The value of ϕ for the Heisenberg model in two and three dimensions is uncertain (Stinchcombe 1983).

In the region $p < p_c$, the correlation length grows as the critical point is approached. The limit to this size is either the size of the percolation cluster or the correlation length of the one-dimensional chain. Birgeneau et al. (1980) point out that this suggests an additive relation in reciprocal space and suggest a form for the correlation length consistent with Eqs. 9.1 and 9.2:

$$[\xi(p - p_c, T)]^{-1} = A(p - p_c)^{\nu_p} + B\xi_1^{-\nu_T} \tag{9.3}$$

9.3. Random Fields

Random fields can have interesting and drastic effects on critical phase transitions. Consider, for instance, an ordered Ising ferromagnet and apply random small fields h_i along z at each site i such that

$$\langle h_i \rangle = 0, \qquad \langle h_i^2 \rangle \neq 0$$

In zero field there is a symmetry-breaking at the critical point and one of two degenerate states has been chosen, with spins either aligned along

+z or along −z. The presence of the random field affects the implementation of the symmetry-breaking process, since in some regions there will be an excess of fields along +z and in other regions along −z. This will affect the nature of the critical fluctuations in a drastic way; for example, the Ginzburg–Landau theory only applies with $d \geq 6$ (Aharony et al. 1976), while we have seen that for the standard models the corresponding number of dimensions is four. It is expected that the presence of random fields will destroy long-range order in two dimensions but not in three (Imry and Ma 1975, Grinstein 1984) for the Ising model. For the X–Y and Heisenberg models, long-range order is predicted to be destroyed in three dimensions.

At first sight, there seems to be no way of applying such random fields to a real system; however, Fishman and Aharony (1979) showed that application of a uniform field along z to a diluted Ising antiferromagnet is a realization of the model. Some local regions will have more spins on spin-up sites than on spin-down sites, so that they will try to form one type of domain, while other local regions will be the other way round. This situation is equivalent to the Ising ferromagnet in a random field and is readily brought about experimentally.

Fishman and Aharony (1979) have applied scaling theory to the random-field Ising model and find

$$G(t, H) = |t|^{2-\alpha} f(H^2/|t|^{\phi_R}) \tag{9.4}$$

where ϕ_R is a crossover exponent (cf. Eq. 8.7) between the diluted Ising model and the random-field Ising model. Aharony (1986) shows that $\phi_R > \gamma$, where γ is the susceptibility exponent for the diluted Ising model. He estimates by renormalization group techniques that

$$\phi_R \simeq 1.1\gamma \tag{9.5}$$

Fishman and Aharony (1979) show that it follows from Eq. 9.4 that

$$T_N(H) = T_N(0) - AH^2 - BH^{2/\phi_R} \tag{9.6}$$

where $T_N(H)$ is the Néel temperature in field H, and A and B are positive constants. The term $-AH^2$ represents the variation of the Néel temperature with field in the pure Ising antiferromagnet; for the condition actually used in the experiments that have been done, this term is small compared with the term involving the constant B.

It also follows from this scaling theory that if the temperature is fixed at $T_N(0)$ and the field H is varied, then

$$\xi \sim H^{-\nu_H} \qquad \nu_H = \frac{\gamma/\phi_R}{1 - \eta/2} \tag{9.7}$$

and, using Eq. 5.36,

$$\hat{C}(\mathbf{0}, 0, H) \sim H^{-\gamma_H} \qquad \gamma_H = 2\gamma/\phi_R \tag{9.8}$$

If the random field destroys the long-range order at all temperatures, then it is possible to extend these equations to all temperatures below T_N and so to obtain critical exponents $v_H(T)$ and $\gamma_H(T)$ that are functions of the temperature.

Suggested Further Reading

Stinchcombe (1983)
Grinstein (1984)

10

BASIC PROPERTIES OF THERMAL NEUTRONS

10.1. Introduction

Nuclear reactors provide a copious source of thermal neutrons. To a reasonable approximation, neutrons produced by nuclear fission are moderated within the reactor to form a gas with a Maxwellian distribution of speeds corresponding to a temperature equal to that of the moderating material. This moderating material is typically light or heavy water at a temperature somewhat above 300 K, though if more or less energetic neutrons are required it is possible to locally heat or cool a part of the moderator and so produce a *hot source* or a *cold source*. Figure 10.1 plots the Maxwellian flux distribution coming from moderators at 320 K and at 25 K. The peak flux for a temperature of 320 K occurs at a wavenumber of 4.5 Å^{-1}, which corresponds to a neutron wavelength of 1.40 Å. By cooling the moderator to 25 K the peak flux is shifted to a wavenumber of 1.25 Å^{-1}, corresponding to a neutron wavelength of 5.0 Å. These wavelengths are of the order of interatomic distances, so that neutrons are well suited to probe properties on an atomic length scale.

Neutrons that have been thermalized in a reactor have energies of the order of magnitude of $k_B T_M$, where T_M is the moderator temperature; thus, if thermal neutrons are used to probe materials for energy levels of this magnitude and slightly lower, the fractional change in energy in a scattering event will be large, and so easy to measure.

For convenience we write down the numerical factors E_m involved in the interconversion of neutron energy in millielectronvolts (meV), neutron energy E_t in terahertz (THz), neutron wavenumber k in reciprocal angstroms (Å^{-1}), neutron wavelength λ in angstroms (Å), and neutron velocity v in kilometers per second. The relevant equations are

$$E_m = 4.136 E_t = 2.072 k^2 = \frac{81.81}{\lambda^2} = 5.227 v^2$$

Almost all the literature of neutron scattering uses Å as the unit of distance and either meV or THz as the unit of energy, so these units will be used throughout this book.

Thermal neutrons interact with matter in two ways. The first of these ways is the interaction with the nucleus via the so-called strong force. This interaction is strong, but extends only over a distance of the order of 10 fm (femtometers), the size of the atomic nucleus. It can have the effect

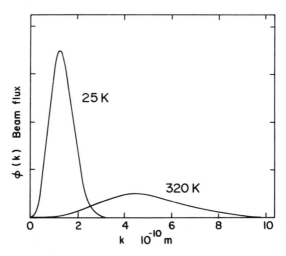

Figure 10.1. Flux distribution in the beam of neutrons from a moderator at 25 K and from a moderator at 320 K. The distributions are normalized to have the same total flux.

of the neutron being scattered by the nucleus or of the neutron being absorbed by the nucleus. Our principal interest in this book is in scattering events, which can, of course, be observed by detecting the scattered neutrons. The scattering nucleus is a bound quantum system with discrete energy levels separated by energies of the order of 0.1 to 1 meV. Since thermal neutrons have energies of tens of millielectronvolts, they will not be able to produce nuclear excitations and the scattering nucleus will remain in the same internal quantum state. Because the interaction only extends over short distances around the nucleus, nuclear scattering is a relatively improbable event and the mean free path of a thermal neutron in condensed matter is of the order of 1 cm.

The second way in which thermal neutrons interact with matter is via magnetic forces. The neutron has a magnetic dipole moment

$$\mu_n = -\gamma \mu_N \sigma \qquad (10.1)$$

where $\gamma = 1.913$, μ_N is the nuclear magneton and σ is a Pauli spin operator for a particle with spin quantum number 1/2. In magnetic solids there are unpaired electrons with magnetic moments that will interact with the neutron's magnetic moment. This interaction is weaker than the nuclear interaction, but extends over much larger distances. The net result of the interaction is to give the thermal neutron a mean free path for magnetic scattering events that is of the same order as the mean free path for nuclear scattering events, that is about 1 cm.

There are other interactions between thermal neutrons and matter (Shull 1967), but these are orders of magnitude smaller and can be neglected except in rather special circumstances.

10.2. The Cross Section

A quantitative discussion of neutron scattering requires the introduction of the concept of the cross section, which is denoted by the symbol σ. The idea is illustrated in Figure 10.2, where a parallel beam of neutrons of flux I_0 neutrons per unit area per second and energy E is incident on a target. We assume that the target area is small compared with that of the beam and that the target is sufficiently small that the probability of scattering of a neutron incident on the target is not large. In view of the fact that the mean free path of thermal neutrons in condensed matter is of the order of 1 cm, this assumption is not particularly restrictive. If the target does scatter some of the incident neutrons, then at a large distance from the target we can detect neutrons that are scattered into a small element of solid angle $d\Omega$ with energy E'. We define the cross section for this scattering by putting the number of neutrons per second scattered into solid angle $d\Omega$ with energy between E' and $E' + dE'$ equal to

$$I_0 \frac{d^2\sigma_s}{d\Omega\, dE'} d\Omega\, dE'$$

The quantity $d^2\sigma_s/d\Omega\, dE'$ is the *partial differential cross section*.

Integrating this over dE', we put the number of neutrons per second scattered into solid angle $d\Omega$ (regardless of energy E') to be

$$I_0\, d\Omega \int_0^\infty \frac{d^2\sigma_s}{d\Omega\, dE'} dE' = I_0 \frac{d\sigma_s}{d\Omega} d\Omega$$

and $d\sigma_s/d\Omega$ is the *differential cross section*. If we now integrate over solid angle, we find that the number of neutrons scattered per second is

$$I_0 \int \frac{d\sigma_s}{d\Omega} d\Omega = I_0 \sigma_s$$

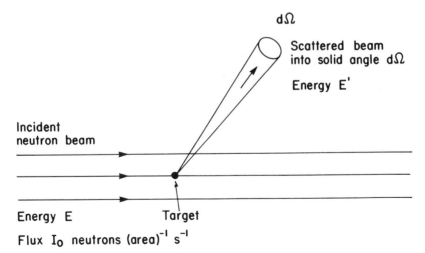

Figure 10.2. The scattering of a neutron beam by a small target into solid angle $d\Omega$.

and σ_s is known as the *scattering cross section*. In a similar way, we can put the number of neutrons that are absorbed by the target per second equal to

$$I_0 \sigma_a$$

and σ_a is known as the *absorption cross section*. Finally we can write down the number of neutrons per second that have their momentum changed in the target as

$$I_0 \sigma_t$$

and σ_t is known as the *total cross section*. Since the only events that normally happen to a neutron are that it is scattered or absorbed, we can write

$$\sigma_t = \sigma_a + \sigma_s$$

The cross section has the dimensions of area and is usually quoted in barns (1 barn = 10^{-28} m^2) for neutron scattering, and in nuclear physics generally. This is about the projected area of an atomic nucleus. If the target consists of many atoms, we can either deal with the cross section for the whole target, or take it per atom or per chemical formula unit.

Cross sections are quantities that are actually measured in experiments, so it is usual to try to cast theoretically expressions for the scattering into the form of cross sections and to compare theory with experiment at this level. To investigate critical phenomena by magnetic neutron scattering, we are going to have to derive theoretical expressions for the magnetic scattering cross section. We will find this cross section to be directly proportional to the correlation function in reciprocal space, so that the scattering measurements give detailed information about the critical fluctuations.

The scattering cross section of the target can be calculated by quantum scattering theory. Since, as we have pointed out, the scattering probability is small, it is sufficient to make the calculation in the first Born approximation. The initial state of the neutron is designated $|\mathbf{k}\sigma\rangle$, where $\mathbf{k}$ is the neutron wavevector and σ labels the spin state, and the final state of the neutron is designated as $|\mathbf{k}'\sigma'\rangle$. The target is initially in quantum state $|\lambda\rangle$ and after the scattering event is in quantum state $|\lambda'\rangle$. The partial differential cross section for scattering from the initial state $|\mathbf{k}\sigma\lambda\rangle$ of the system to the final state $|\mathbf{k}'\sigma'\lambda'\rangle$ is given by (Squires 1978, Lovesey 1984)

$$\frac{d^2\sigma_s}{d\Omega \, dE'} = \frac{k'}{k}\left(\frac{m}{2\pi\hbar^2}\right)^2 |\langle \mathbf{k}'\sigma'\lambda'| V |\mathbf{k}\sigma\lambda\rangle|^2 \, \delta(E_\lambda - E_{\lambda'} + E - E') \quad (10.2)$$

where E_λ is the energy of the target in state $|\lambda\rangle$, m is the mass of the neutron, and V is the interaction potential between the neutron and the target.

In order to proceed, we need to know the interaction potential V. We consider two cases, first interactions with the nucleus via the nuclear force, and second magnetic interactions.

10.3. Nuclear Scattering

For nuclear scattering we describe the interaction potential by a quantity known as the *Fermi pseudopotential*

$$V = \frac{2\pi\hbar^2}{m} b\, \delta(\mathbf{r} - \mathbf{R}) \qquad (10.3)$$

where $\mathbf{R}$ is the nuclear position and $\mathbf{r}$ is the neutron's position. The variable b has the dimensions of length and is known as the *scattering length*. It is typically found to be of the order of 10 fm.

The Fermi pseudopotential cannot be the correct interaction potential, since the interaction is not actually infinite anywhere. However, the nuclear size and the range of the nuclear force are so small compared with the wavelength of a thermal neutron that the delta function pseudopotential gives an excellent representation of the actual scattering.

We are now in a position to evaluate the matrix elements of V over the neutron spacial coordinates. The incident neutron wavefunction will vary as $\exp(i\mathbf{k}\cdot\mathbf{r})$ and the scattered neutron wavefunction will vary as $\exp(i\mathbf{k}'\cdot\mathbf{r})$. The matrix element for scattering from a fixed nucleus at position $\mathbf{R}$ is

$$\langle \mathbf{k}'|\, V\, |\mathbf{k}\rangle \sim \int \exp(-i\mathbf{k}'\cdot\mathbf{r})\, \delta(\mathbf{r}-\mathbf{R})\, \exp(i\mathbf{k}\cdot\mathbf{r})\, d\mathbf{r} \qquad (10.4)$$

$$\langle \mathbf{k}'|\, V\, |\mathbf{k}\rangle \sim \exp(i\boldsymbol{\kappa}\cdot\mathbf{R}) \qquad (10.5)$$

with $\boldsymbol{\kappa} = \mathbf{k} - \mathbf{k}'$: $\boldsymbol{\kappa}$ is known as the *scattering vector*. The momentum transfer in the scattering event is $\hbar\boldsymbol{\kappa}$, since the initial and final momenta of the neutron are $\hbar\mathbf{k}$ and $\hbar\mathbf{k}'$, respectively. We can now put this result into Eq. 10.2 for the scattering cross section. We must take care to use normalized neutron wavefunctions at this stage and we will assume that the scattering is from an assembly of fixed nuclei at positions $\mathbf{R}_j$ that have scattering length b_j. We will further assume that this scattering length is independent of the neutron spin state $|\sigma\rangle$. We can then sum over all such spin states to find (Squires 1978; Lovesey 1984)

$$\frac{d^2\sigma_s}{d\Omega\, dE'} = \frac{k'}{k} \left| \langle \lambda'| \sum_j b_j \exp(i\boldsymbol{\kappa}\cdot\mathbf{R}_j) |\lambda\rangle \right|^2 \delta(E_\lambda - E_{\lambda'} + E - E') \qquad (10.6)$$

If the scattering is from a single atom that is held fixed, so that its quantum state does not change,

$$|\lambda\rangle = |\lambda'\rangle$$

then $E_\lambda = E_{\lambda'}$ and

$$\frac{d^2\sigma_s}{d\Omega\, dE'} = \frac{k'}{k} b^2\, \delta(E - E')$$

This scattering is zero unless $E = E'$, so that the δ function is nonzero; then $k = k'$ and

$$\frac{d\sigma_s}{d\Omega} = \int b^2 \, \delta(E - E') \, dE'$$
$$= b^2$$

It follows that

$$\sigma_s = \int \frac{d\sigma_s}{d\Omega} d\Omega = 4\pi b^2 \tag{10.7}$$

The scattering cross section from a single fixed atom is $4\pi b^2$.

10.4. Separation of the Nuclear Scattering into Coherent and Incoherent Parts

In this section we consider the consequences of the fact that the nuclear scattering length b of an atom depends not just on the chemical nature of the atom but also on the nature of the nucleus of the atom. If a particular element has more than one isotope, the scattering length of each isotope will be different. In addition to this isotope effect, the scattering of a particular nucleus will also depend on the relative orientation of the spin of the neutron and of the nucleus, because the nuclear force is dependent on the relative spin orientation. For a nucleus with nonzero spin, this will have the effect of making the scattering length not always the same, but dependent on the spin orientation.

Both these effects have the effect of making the scattering length vary from site to site, even for sites occupied entirely by atoms of the same element. The matrix element squared term of Eq. 10.6 can be written, by expanding the square and realizing that b_j is independent of the quantum state (λ) of the scattering system,

$$\left| \langle \lambda' | \sum_j b_j \exp(i\boldsymbol{\kappa} \cdot \mathbf{R}_j) | \lambda \rangle \right|^2 = \sum_{jl} b_j b_l F_j F_l^* \tag{10.8}$$

where we have assumed b to be real and

$$F_j = \langle \lambda' | \exp(i\boldsymbol{\kappa} \cdot \mathbf{R}_j) | \lambda \rangle$$

It will normally be the case that the isotopic distribution is random, so that, if the nuclear spins have random orientation, then the value of the scattering length b_j at the jth site will not be correlated with j. This means that in Eq. 10.8, b_j and b_l can be replaced by their average values $\bar{b}$, except in the special case when j and l are the same, when the term $b_j b_l \, (= b_j^2)$ must be replaced by the mean square value of b, which is $\overline{b^2}$. Thus,

$$\sum_{jl} b_j b_l F_j F_l^* = {\sum_{jl}}' (\bar{b})^2 F_j F_l^* + \sum_j \overline{b^2} F_j F_j^*$$

where the prime on the first sum on the right-hand-side means that the

Basic Properties of Thermal Neutrons

term $j = l$ is excluded from the summation. It is convenient to rewrite the primed summation as an unrestricted summation by adding $\sum_j (\bar{b})^2 F_j F_j^*$ to the first term and subtracting it from the second term. This gives

$$\sum_{jl} b_j b_l F_j F_l^* = (\bar{b})^2 \sum_{jl} F_j F_l^* + (\overline{b^2} - (\bar{b})^2) \sum_j F_j F_j \quad (10.9)$$

The first of these terms is known as the *coherent term* and the second as the *incoherent term*. The scattering can be divided into two parts, which are known as the coherent and the incoherent scattering. If the cross sections for these parts are σ_c and σ_i, then for a single fixed atom, analogy with Eq. 10.7 shows that

$$\sigma_c = 4\pi (\bar{b})^2$$
$$\sigma_i = 4\pi (\overline{b^2} - (\bar{b})^2) \quad (10.10)$$

and

$$\sigma_s = \sigma_c + \sigma_i$$

These cross sections have been measured experimentally and are tabulated in the literature for the elements and many isotopes (Sears 1984). The Appendix of this book lists these values in commonly occuring cases, together with the absorption cross section.

10.5. Magnetic Scattering

This scattering arises from the interaction of the neutron magnetic moment $\boldsymbol{\mu}_n$ with the local magnetic field **B**. The local field due to an electron at position **R** with momentum **p** and magnetic moment $\boldsymbol{\mu}_e$ arising from the electron spin is

$$\mathbf{B} = \frac{\mu_0}{4\pi} \left[\left(\text{curl} \frac{\boldsymbol{\mu}_e \times \mathbf{R}}{|\mathbf{R}|^3} \right) - \frac{2\mu_B}{\hbar} \frac{\mathbf{p} \times \mathbf{R}}{|\mathbf{R}|^3} \right] \quad (10.11)$$

where the first term is the field due to the electron's spin and the second is the field due to the electron's orbital motion. The interaction potential with the neutron is given by

$$V = -\boldsymbol{\mu}_n \cdot \mathbf{B} \quad (10.12)$$

To calculate the magnetic scattering cross section, we must substitute this potential V into Eq. 10.2 and evaluate the matrix elements $\langle k'| V |k \rangle$. The reader is referred to Squires (1978) or to Lovesey (1984) for the details of this calculation; the result is that the magnetic scattering is given by

$$\frac{d^2 \sigma_s}{d\Omega \, dE'} = (\gamma r_0)^2 \frac{k'}{k} |\langle \sigma' \lambda'| \boldsymbol{\sigma} \cdot \mathbf{Q}_\perp |\sigma \lambda \rangle|^2 \, \delta(E_\lambda - E_{\lambda'} + E - E') \quad (10.13)$$

where r_0 is known as the classical radius of the electron, given by

$$r_0 = \frac{\mu_0}{4\pi} \frac{e^2}{m_e}$$
$$= 2.818 \text{ fm} \tag{10.14}$$

and $\mathbf{Q}_\perp$ is the projection of a vector $\mathbf{Q}$ onto the plane perpendicular to the scattering vector. $\mathbf{Q}_\perp$ is given by the equation

$$\mathbf{Q}_\perp = \sum_l |\kappa|^{-2} \exp(i\boldsymbol{\kappa} \cdot \mathbf{r}_l) \left[\boldsymbol{\kappa} \times (\mathbf{s}_l \times \boldsymbol{\kappa}) + \frac{i}{\hbar}(\mathbf{p}_l \times \boldsymbol{\kappa}) \right] \tag{10.15}$$

with

$$\boldsymbol{\mu}_e = -2\mu_B \mathbf{s} \tag{10.16}$$

where $\mu_B (=e\hbar/(2m_e))$ is the Bohr magneton, m_e is the mass of the electron and $\mathbf{s}$ is the Pauli spin operator for the electron. The summation is over the l electrons of the scattering system.

The vector $\mathbf{Q}$, whose projection $\mathbf{Q}_\perp$ appears in the cross section, Eq. 10.13, is related to the Fourier transform of the magnetization density $\mathbf{M}(\mathbf{r})$ at $\mathbf{r}$ by the equations (Squires 1978, Lovesey 1984)

$$2\mu_B \mathbf{Q} = \mathbf{M}(\boldsymbol{\kappa}) \tag{10.17}$$

and

$$\mathbf{M}(\boldsymbol{\kappa}) = \int \mathbf{M}(\mathbf{r}) \exp(i\boldsymbol{\kappa} \cdot \mathbf{r}) \, d\mathbf{r} \tag{10.18}$$

The magnetization arises from both the electron's spin and from its orbital motion. It is the component of the magnetization perpendicular to the scattering vector $\boldsymbol{\kappa}$ that gives the neutron scattering through its Fourier transform.

The relationship between $\mathbf{Q}$ and $\mathbf{Q}_\perp$ such that $\mathbf{Q}_\perp$ is the projection of $\mathbf{Q}$ in the direction perpendicular to $\boldsymbol{\kappa}$ is

$$\mathbf{Q}_\perp = |\kappa|^{-2} \boldsymbol{\kappa} \times (\mathbf{Q} \times \boldsymbol{\kappa}) \tag{10.19}$$

An alternative formulation of this equation can be written down by realizing that

$$\mathbf{Q}_\| = |\kappa|^{-2} (\mathbf{Q} \cdot \boldsymbol{\kappa}) \boldsymbol{\kappa}$$

so that

$$\mathbf{Q}_\perp = \mathbf{Q} - \mathbf{Q}_\|$$
$$= \mathbf{Q} - |\kappa^{-2}| (\mathbf{Q} \cdot \boldsymbol{\kappa}) \boldsymbol{\kappa} \tag{10.20}$$

10.6. Spin-only Scattering from Unpolarized Neutrons

If the neutron beam is unpolarized, the initial neutron state $|\sigma\rangle$ will be random and uncorrelated with $\mathbf{Q}$ or with the initial state of the scatterer $|\lambda\rangle$. Then

$$\langle \sigma'\lambda' | \boldsymbol{\sigma} \cdot \mathbf{Q}_\perp | \sigma\lambda \rangle = \sum_\alpha \langle \sigma' | \sigma_\alpha | \sigma \rangle \langle \lambda' | Q_{\perp\alpha} | \lambda \rangle \tag{10.21}$$

It is often convenient to rewrite this equation in terms of **Q** rather than of $\mathbf{Q}_\perp$. When this is done, the cross section for unpolarized neutron scattering from state $|\lambda\rangle$ to state $|\lambda'\rangle$ becomes

$$\frac{d^2\sigma_s}{d\Omega\, dE'} = (\gamma r_0)^2 \frac{k'}{k} \sum_{\alpha\beta} (\delta_{\alpha\beta} - \hat{\kappa}_\alpha \hat{\kappa}_\beta) \langle \lambda | \mathbf{Q}_\alpha^+ | \lambda' \rangle$$
$$\times \langle \lambda' | \mathbf{Q}_\beta | \lambda \rangle\, \delta(E_\lambda - E_{\lambda'} + E - E') \quad (10.22)$$

where $\hat{\kappa}_\alpha$ is the direction cosine of **κ** along the α axis.

Equation 10.22 is particularly useful because, through Eqs. 10.17 and 10.18 we can relate the scattering directly to the magnetization density **M(r)**. We will now proceed by making the assumption that the magnetization density can be separated into distinct parts corresponding to each atomic site j at position R_j. This assumption seems quite reasonable for localized magnetic electrons whose wave function is very small between the atoms, but for delocalized electrons there is a problem in defining the spatial extent of an atomic site. We use this assumption to write

$$\mathbf{M}(\mathbf{R}) = \sum_j \mathbf{M}(\mathbf{R}_j + \mathbf{r}_j) \quad (10.23)$$

where $\mathbf{r}_j$ is a vector within the atomic site centered on $\mathbf{R}_j$, so that each position vector **r** where the magnetization density is nonzero can be assigned uniquely to one atomic site. We combine Eqs. 10.17, 10.18, and 10.23 to get

$$2\mu_B \mathbf{Q} = \int \sum_j \mathbf{M}(\mathbf{R}_j + \mathbf{r}_j) \exp(i\boldsymbol{\kappa} \cdot \mathbf{R}_j) \exp(i\boldsymbol{\kappa} \cdot \mathbf{r}_j)\, d\mathbf{r}$$

Each volume element $d\mathbf{r}$, where **M(r)** is nonzero, corresponds to a particular vector $\mathbf{r}_j$ and we can replace $d\mathbf{r}$ by $d\mathbf{r}_j$ and still do the integral over all volume elements:

$$2\mu_B \mathbf{Q} = \sum_j \exp(i\boldsymbol{\kappa} \cdot \mathbf{R}_j) \int \mathbf{M}(\mathbf{R}_j + \mathbf{r}_j) \exp(i\boldsymbol{\kappa} \cdot \mathbf{r}_j)\, d\mathbf{r}_j \quad (10.24)$$

where the integral is now to be taken over each atomic site. The site integral in Eq. 10.24 will be especially simple if the magnetization vector is in the same direction throughout the atomic site. This is likely to be the case if the atom has no orbital angular momentum and the magnetism arises from spin only. Then we say that the jth site, centered at $\mathbf{R}_j$, has a net magnetic moment $\mathbf{M}_j$ and a spin $\mathbf{S}_j$ given by

$$\mathbf{M}_j = 2\mu_B \mathbf{S}_j = \int \mathbf{M}(\mathbf{R}_j + \mathbf{r}_j)\, d\mathbf{r}_j \quad (10.25)$$

A quantity f_j known as the *magnetic form factor* can be defined by

$$f_j(\boldsymbol{\kappa}) = \frac{\int \mathbf{M}(\mathbf{R}_j + \mathbf{r}_j) \exp(i\boldsymbol{\kappa} \cdot \mathbf{r}_j)\, d\mathbf{r}_j}{\int \mathbf{M}(\mathbf{R}_j + \mathbf{r}_j)\, d\mathbf{r}_j} \quad (10.26)$$

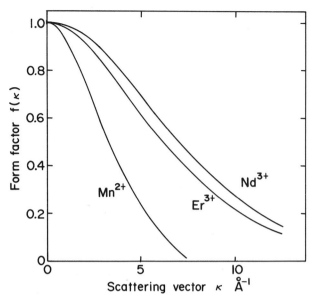

Figure 10.3. Magnetic form factors for Mn^{2+}, Er^{3+}, and Nd^{3+} ions, from calculations of Watson and Freeman (1961) and Blume et al. (1962, 1964).

The magnetic form factor is the Fourier transform of the atomic magnetization density. It is closely analogous to the atomic form factor used in describing X-ray scattering, which is the Fourier transform of the charge density. Figure 10.3 shows typical magnetic form factors for Mn^{2+}, Er^{3+}, and Nd^{3+} ions. Mn^{2+} corresponds to a spin-only moment with no orbital moment. In Er^{3+} there are both spin and orbital moments aligned parallel to each other, while in Nd^{3+} the alignment is antiparallel. The plots correspond to magnetic form factors as calculated by Watson and Freeman (1961) and by Blume et al. (1962, 1964) from Hartree–Fock calculations of the magnetization density. It is apparent that the magnetic form factor falls off sharply as the scattering vector increases; because of this, the magnetic scattering cross section becomes small for large scattering vectors.

Incorporation of Eqs. 10.25 and 10.26 into Eq. 10.24 gives, for spin only magnetization,

$$\mathbf{Q} = \sum_j \mathbf{S}_j f_j(\boldsymbol{\kappa}) \exp(i\mathbf{k} \cdot \mathbf{R}_j) \tag{10.27}$$

We can now write the cross section (Eq. 10.22) for spin-only scattering from unpolarized neutrons as

$$\frac{d^2\sigma_s}{d\Omega\, dE'} = (\gamma r_0)^2 \frac{k'}{k} \sum_{\alpha\beta} \Bigg[(\delta_{\alpha\beta} - \hat{\kappa}_\alpha \hat{\kappa}_\beta) \sum_{jl} \exp(i\mathbf{k} \cdot (\mathbf{R}_j - \mathbf{R}_l))$$
$$\times \langle \lambda | S_{l\alpha} f_l(\boldsymbol{\kappa}) | \lambda' \rangle \langle \lambda' | S_{j\beta} f_j(\boldsymbol{\kappa}) | \lambda \rangle \Bigg]$$
$$\times \delta(E_\lambda - E_{\lambda'} + E - E') \tag{10.28}$$

where we have used the fact that, since $S_{l\alpha}$ corresponds to an observable quantity, it must be a Hermitian operators with $S_{l\alpha}^{\dagger} = S_{l\alpha}$.

Suggested Further Reading

Bacon (1975)
Lovesey (1984)
Squires (1978)

11

CORRELATION FUNCTION FORMALISM

11.1. Nuclear Scattering

In Chapter 10 we wrote down the basic equations for the neutron-scattering cross section. We made no contact, however, with the earlier part of the book about critical phenomena. This chapter will establish such contact and show that the neutron-scattering cross section is closely related to the correlation functions that were defined in Section 5.3. Because the formalism is somewhat less cumbersome, we start with the nuclear correlation function. Suppose that the scattering system is a set of N nuclei that all have the same scattering length b. Then we can write Eq. 10.6 for scattering from the set of states $|\lambda\rangle$ to the set of states $|\lambda'\rangle$ as

$$\frac{d^2\sigma}{d\Omega\, dE'} = \frac{k'}{k} Nb^2 S(\mathbf{\kappa}, \omega) \tag{11.1}$$

with

$$S(\mathbf{\kappa}, \omega) = N^{-1} \sum_\lambda p_\lambda \sum_{\lambda'} \left| \langle \lambda' | \sum_j \exp(i\mathbf{\kappa} \cdot \mathbf{R}_j) | \lambda \rangle \right|^2 \delta(E_\lambda - E_{\lambda'} + \hbar\omega) \tag{11.2}$$

where p_λ is the probability of the scatterer being initially in the state λ and the energy transfered from the neutron to the scatterer is given by

$$\hbar\omega = E - E' \tag{11.3}$$

$S(\mathbf{\kappa}, \omega)$ is known as the *scattering function*.
We now make use of the following relations:

$$\delta(\hbar\omega + c) = \hbar^{-1} \delta(\omega + c/\hbar)$$

$$= \hbar^{-1} (2\pi)^{-1} \int_{-\infty}^{\infty} \exp[i(\omega + c/\hbar)\tau]\, d\tau \tag{11.4}$$

where τ is the time, and write

$$S(\mathbf{\kappa}, \omega) = (hN)^{-1} \sum_\lambda p_\lambda \sum_{\lambda'} \int_{-\infty}^{\infty} \exp(i\omega\tau)\, d\tau \sum_{jl} \langle \lambda | \exp(-i\mathbf{\kappa} \cdot \mathbf{R}_l) | \lambda' \rangle$$

$$\times \langle \lambda' | \exp(-iE_{\lambda'}\tau/\hbar) \exp(i\mathbf{\kappa} \cdot \mathbf{R}_j) \exp(iE_\lambda \tau/\hbar) | \lambda \rangle \tag{11.5}$$

Now, Schrödinger's equation shows that

$$\mathcal{H} |\lambda\rangle = E_\lambda |\lambda\rangle \quad \text{and} \quad \langle \lambda' | \mathcal{H} = \langle \lambda' | E_{\lambda'}$$

and it follows that

$$\mathcal{H}^n |\lambda\rangle = E_\lambda^n |\lambda\rangle \quad \text{and} \quad \langle \lambda'| \mathcal{H}^n = \langle \lambda'| E_{\lambda'}^n$$

by operating n times with $\mathcal{H}$ on $|\lambda\rangle$ or on $\langle \lambda'|$. Since both $\exp(iE_\lambda \tau/\hbar)$ and $\exp(i\mathcal{H}\tau/\hbar)$ can be expanded as power series, we find that

$$\exp(iE_\lambda \tau/\hbar) |\lambda\rangle = \exp(i\mathcal{H}\tau/\hbar) |\lambda\rangle$$

and

$$\langle \lambda'| \exp(-iE_{\lambda'} \tau/\hbar) = \langle \lambda'| \exp(-i\mathcal{H}\tau/\hbar)$$

so that the last matrix element in Eq. 11.5 becomes

$$\langle \lambda'| \exp(-i\mathcal{H}\tau/\hbar) \exp(i\boldsymbol{\kappa} \cdot \mathbf{R}_j) \exp(i\mathcal{H}\tau/\hbar) |\lambda\rangle \tag{11.6}$$

Time-dependent properties in quantum mechanics can be represented either in the Schrödinger representation, where the states are time dependent and the operators are time independent, or in the Heisenberg representation, where the states are time independent and the operators are time dependent. If we choose to use the Heisenberg representation, the time dependence of an operators $\mathbf{O}(\tau)$ is given by

$$\mathbf{O}(\tau) = \exp(-i\mathcal{H}\tau/\hbar)\, \mathbf{O} \exp(i\mathcal{H}\tau/\hbar)$$

where $\mathbf{O}$ is the time-independent operator. The matrix element in expression 11.6 becomes

$$\langle \lambda'| \exp(i\boldsymbol{\kappa} \cdot \mathbf{R}_j(\tau)) |\lambda\rangle$$

and Eq. 11.5 becomes

$$S(\boldsymbol{\kappa}, \omega) = (hN)^{-1} \sum_\lambda p_\lambda \sum_{\lambda'} \int_{-\infty}^{\infty} \exp(i\omega\tau)\, d\tau$$
$$\times \sum_{jl} \langle \lambda| \exp(-i\boldsymbol{\kappa} \cdot \mathbf{R}_l(0)) |\lambda'\rangle$$
$$\times \langle \lambda'| \exp(i\boldsymbol{\kappa} \cdot \mathbf{R}_j(\tau)) |\lambda\rangle$$

Now we can use the results

$$\sum_{\lambda'} |\lambda'\rangle \langle \lambda'| = 1 \tag{11.7}$$

and

$$\sum_\lambda p_\lambda \langle \lambda| \mathbf{O} |\lambda\rangle = \langle \mathbf{O} \rangle \tag{11.8}$$

where $\mathbf{O}$ is any operator, to get

$$S(\boldsymbol{\kappa}, \omega) = (hN)^{-1} \sum_{jl} \int_{-\infty}^{\infty} \exp(i\omega\tau)\, d\tau$$
$$\times \langle \exp(-i\boldsymbol{\kappa} \cdot \mathbf{R}_l(0)) \exp(i\boldsymbol{\kappa} \cdot \mathbf{R}_j(\tau)) \rangle \tag{11.9}$$

This result was first obtained by van Hove (1954a). It expresses the cross section as the temporal Fourier transform of a correlation function between the position vector of the lth atom at time zero and the jth atom at time τ.

If we now go back and remove the assumption that all the nuclei have the same scattering length b, we can see from Section 10.4 that the scattering will separate into a coherent and an incoherent part. The coherent part will be given by Eqs. 11.1 and 11.9 with b^2 in Eq. 11.1 replaced by $(\bar{b})^2$. The incoherent part will have b^2 in Eq. 11.1 replaced by $\overline{b^2} - (\bar{b})^2$; Eq. 11.9 will then be restricted to the case $l = j$ and will give a so-called self correlation function that correlates the position vector of an atom at time zero and the position vector of the same atom at time τ.

11.2. Magnetic Scattering

In this section we go through the same steps for the magnetic scattering as we did for the nuclear scattering in the previous section. We simplify by assuming the magnetic scattering to come from N identical atoms with form factor $f(\kappa)$ that is the same for all states $|\lambda\rangle$ that give rise to significant scattering. Then it is conventional to define a magnetic scattering function $S^{\alpha\beta}(\kappa, \omega)$ in an analogous way to the nuclear scattering function (Eqs. 11.1 and 11.2), to give (from Eq. 10.28)

$$\frac{d^2\sigma}{d\Omega\, dE'} = \frac{k'}{k}\frac{N}{\hbar}(\gamma r_0)^2 |f(\kappa)|^2 \sum_{\alpha\beta} (\delta_{\alpha\beta} - \hat{k}_\alpha \hat{k}_\beta) S^{\alpha\beta}(\kappa, \omega) \quad (11.10)$$

with

$$S^{\alpha\beta}(\kappa, \omega) = \hbar N^{-1} \sum_\lambda p_\lambda \sum_{\lambda'} \sum_{jl} \exp(i\kappa(\mathbf{R}_j - \mathbf{R}_l))$$
$$\times \langle \lambda | S_{l\alpha} | \lambda' \rangle \langle \lambda' | S_{j\beta} | \lambda \rangle \, \delta(E_\lambda - E_{\lambda'} + \hbar\omega) \quad (11.11)$$

The conventional definition of the magnetic scattering function differs from that of the nuclear function by a factor of $\hbar$. The length $\gamma r_0 = 5.391$ fm is analogous to the nuclear scattering length b and is of the same order of magnitude (cf. Appendix). In addition, the magnetic scattering function is expressed as a second-rank tensor as a way of reflecting the physical point that it is only the magnetization in the plane perpendicular to κ that gives rise to neutron scattering.

We go through exactly the same steps we went through in deriving Eq. 11.9 from Eq. 11.2. If we assume that the atomic positions are fixed, we find, following van Hove (1954b), that

$$S^{\alpha\beta}(\kappa, \omega) = (2\pi N)^{-1} \sum_{jl} \exp(i\kappa \cdot (\mathbf{R}_j - \mathbf{R}_l))$$
$$\times \int_{-\infty}^{\infty} \exp(i\omega\tau) \langle S_{l\alpha}(0) S_{j\beta}(\tau) \rangle \, d\tau \quad (11.12)$$

The idea of the frequency-dependent spin correlation function was introduced in Eq. 7.2 and the necessity in the general case for expressing this correlation function in tensor form was pointed out in Eq. 7.10. In fact, the function $\hat{C}$ that we used in Chapters 5 to 7 is the same as the function S that we use to describe the neutron scattering properties. In defining $\hat{C}$ (and C) we assumed all atoms to be equivalent and gave atom j the position vector $\mathbf{0}$ and atom l the position vector $\mathbf{R}$. In this notation

$$S^{\alpha\beta}(\mathbf{\kappa}, \omega) = \hat{C}^{\alpha\beta}(\mathbf{\kappa}, t, \omega)$$

$$= (2\pi)^{-1} \sum_{\mathbf{R}} \exp(i\mathbf{\kappa} \cdot \mathbf{R}) \int_{-\infty}^{\infty} \exp(i\omega\tau) \langle S_{0\alpha}(0) S_{\mathbf{R}\beta}(\tau) \rangle \, d\tau \quad (11.13)$$

It is clear that neutron scattering is going to be able to provide, at least in principle, the detailed information needed to check dynamic scaling. In the next section we show how static correlation functions may be extracted from the scattering function.

11.3. Measure of Static Correlation Functions; The Static Approximation

Suppose we integrate the magnetic correlation function, as defined by Eq. 11.11, over all frequencies ω keeping the scattering vector $\mathbf{\kappa}$ fixed. Then the only neutron variable is in the delta function and the integral is evaluated as

$$\int_{-\infty}^{\infty} \delta(E_\lambda - E_{\lambda'} + \hbar\omega) \, d\omega = \hbar^{-1} \quad (11.14)$$

so that

$$\int_{-\infty}^{\infty} S^{\alpha\beta}(\mathbf{\kappa}, \omega) \, d\omega = N^{-1} \sum_{jl} \exp(i\mathbf{\kappa}(\mathbf{R}_j - \mathbf{R}_l)) \langle S_{l\alpha} S_{j\beta} \rangle \quad (11.15)$$

$$= \sum_{\mathbf{R}} \exp(i\mathbf{\kappa} \cdot \mathbf{R}) \langle S_{0\alpha} S_{\mathbf{R}\beta} \rangle \quad (11.16)$$

where we have used Eqs. 11.7 and 11.8 in deriving Eq. 11.15. In going from Eq. 11.15 to Eq. 11.16 we have used the same notation change as we used in going from Eq. 11.12 to Eq. 11.13.

Except when $\mathbf{\kappa}$ is a reciprocal lattice point, Eq. 11.16 gives the static correlation function of Eq. 5.30 generalized to tensor form. That is

$$C^{\alpha\beta}(\mathbf{\kappa}, t, h) = \int_{-\infty}^{\infty} S^{\alpha\beta}(\mathbf{\kappa}, \omega) \, d\omega \quad (11.17)$$

Now, as we shall see later, it is quite possible to measure the neutron-scattering cross section as a function of ω keeping $\mathbf{\kappa}$ fixed, so we can derive the static correlation function by integration over ω of the measured magnetic scattering function. There is a problem with this process, however, in that the measurements cannot, in practice, be taken over an infinite range of ω, so that we cannot evaluate the contribution to the integral for all values of ω. Accurate measurements can only be taken if there is a good signal-to-noise ratio and if the weight of the integral is concentrated over a relatively narrow range of ω. There is reason to hope that this latter condition might be reasonably satisfied near the critical point, since it was shown in Chapter 7 that the characteristic frequency ω_c tends to zero at the critical point.

There is a second way in which we can attempt to extract the static correlation function from neutron-scattering measurements. This involves measuring the differential magnetic cross section rather than the partial differential magnetic cross section; that is, we measure all the neutrons scattered into solid angle $d\Omega$ without regard to energy. This is easy to do in practice since it just involves using a detector with angular size $d\Omega$. What we will observe will be the integral of the right-hand side of Eq. 11.10 over the scattered neutron energy dE' at fixed scattering angle:

$$\frac{d\sigma}{d\Omega} = \frac{N(\gamma r_0)^2}{k\hbar} \int_0^\infty dE' \, k' \, |f(\kappa)|^2 \sum_{\alpha\beta} (\delta_{\alpha\beta} - \hat{k}_\alpha \hat{k}_\beta) S^{\alpha\beta}(\kappa, \omega) \quad (11.18)$$

This integral involves variables k', κ, and ω that vary with E' according to the equations

$$E' = \frac{\hbar^2 (k')^2}{2m} \quad (11.19)$$

$$E' = E - \hbar\omega \quad (11.20)$$

and

$$\kappa = \mathbf{k} - \mathbf{k}' \quad (11.21)$$

with the direction of $\mathbf{k}'$ fixed and its magnitude varying according to Eq. 11.19. Clearly we cannot, in general, evaluate this integral unless we know the scattering function, and this is what we are trying to measure! There is one circumstance, however, in which matters simplify: this is if all the weight of the scattering function $S^{\alpha\beta}(\kappa, \omega)$ is at low frequencies such that

$$\hbar\omega \ll E \quad (11.22)$$

If this is the case, it will be an excellent approximation to evaluate the integral with k' and κ held constant (such that $|k'| = |k|$). This then gives

$$\frac{d\sigma}{d\Omega} = \frac{N}{\hbar} (\gamma r_0)^2 |f(\kappa)|^2 \sum_{\alpha\beta} (\delta_{\alpha\beta} - \hat{k}_\alpha \hat{k}_\beta) \hat{C}^{\alpha\beta}(\kappa, t, h) \quad (11.23)$$

which enables the spin correlation function $\hat{C}$ to be measured. The approximation that was made in putting $\hbar\omega \ll E$ for all ω that contributes significantly to the scattering in the correlation function is known as the *static approximation*, so called because it gives the "static" correlation function. It can be shown that the approximation is equivalent to assuming that the spin vectors **S** have not had the time to change during the time that it takes the neutron to cross an atom, so that the diffraction pattern corresponds to a "static" set of spins.

The advantage of measuring spin correlation functions in this way is that the differential cross section can be measured more easily and more accurately than the partial differential cross section. The disadvantage is that we do not normally know how good the static approximation is. It might be expected, at first glance, to be quite satisfactory in the critical region, since the characteristic frequency tends to zero at the critical point; however, on second thought we might be a little less sanguine because we realize that in the critical region both the static correlation function and the characteristic frequency vary rapidly as a function of wavevector (cf. Eqs. 5.34, 5.35, 5.36, 7.7, and 7.8), so that even rather small changes of κ due to weight in the integral at low frequencies may change the cross section. In practical application we would use incident neutrons on the high-speed side of the Maxwellian, take a theoretical model for the frequency dependence of the scattering function, and correct iteratively for the static correlations (including a correction for the variation of counter efficiency with E'). So long as the corrections are small, the data should give good measurements of the static correlations.

11.4. Elastic Scattering

Let us go back to the magnetic scattering function as defined by Eq. 11.12 or by Eq. 11.13. This scattering function is a Fourier transform in time and space of a spin correlation function $\langle S_{0\alpha}(0) S_{\mathbf{R}\beta}(\tau)\rangle$ between spins at different times, 0 and τ. Now, in the paramagnetic state all spin correlations will become zero after long times; that is for all α and β and for all **0** and **R**.

$$\lim_{t \to \infty} \langle S_{0\alpha}(0) S_{\mathbf{R}\beta}(\tau)\rangle = 0$$

In the ordered magnetic state this will no longer be true: the ordered state remains ordered for all time so that spin correlations remain finite as $\tau \to \infty$ and there will be at least one value of α and β for any sites **0** and **R** where

$$\lim_{\tau \to \infty} \langle S_{0\alpha}(0) S_{\mathbf{R}\beta}(\tau)\rangle \neq 0$$

Let us divide the spin correlation function into a part that corresponds to $\tau = \infty$ and a remainder, such that

$$\langle S_{0\alpha}(0) S_{\mathbf{R}\beta}(\tau)\rangle = \langle S_{0\alpha}(0) S_{\mathbf{R}\beta}(\infty)\rangle + \langle S_{0\alpha}(0) S'_{\mathbf{R}\beta}(\tau)\rangle \qquad (11.24)$$

where the prime indicates the difference function $S_{\mathbf{R}\beta}(\tau) - S_{\mathbf{R}\beta}(\infty)$.

Fourier transformation of the first of these terms over time gives

$$\int_{-\infty}^{\infty} \exp(i\omega\tau) \langle S_{0\alpha}(0) S_{\mathbf{R}\beta}(\infty)\rangle \, d\tau = 2\pi \delta(\omega) \langle S_{0\alpha}(0) S_{\mathbf{R}\beta}(\infty)\rangle$$

There is a delta function in the scattering from ordered magnetic systems corresponding to zero energy transfer. This scattering is known as the *elastic scattering*. If we use the subscript *el* to refer to elastic scattering, Eq. 11.13 becomes

$$S_{el}^{\alpha\beta}(\mathbf{\kappa}, \omega) = \delta(\omega) \sum_{\mathbf{R}} \exp(i\mathbf{\kappa} \cdot \mathbf{R}) \langle S_{0\alpha}(0) S_{\mathbf{R}\beta}(\infty) \rangle \qquad (11.25)$$

The magnetic ordering is infinite in spatial extent, so that in a crystalline lattice the sum over $\mathbf{R}$ will give rise to delta functions in the scattering vector $\mathbf{\kappa}$ at magnetic reciprocal lattice points. Such scattering is known as *Bragg scattering*.

Let us demonstrate this by taking the example of an ordered ferromagnet with all atoms having the same spin S aligned in the z direction. Then

$$S_{el}^{\alpha\beta}(\mathbf{\kappa}, \omega) = \delta(\omega)\delta_{\alpha z}\delta_{\beta z} \sum_{\mathbf{R}} \exp(i\mathbf{\kappa} \cdot \mathbf{R})(S_z)^2$$

$$= \delta(\omega)\delta_{\alpha z}\delta_{\beta z}(S_z)^2 \frac{(2\pi)^3}{v_0} \sum_{\mathbf{g}} \delta(\mathbf{\kappa} - \mathbf{g}) \qquad (11.26)$$

where v_0 is the volume of the unit cell and $\mathbf{g}$ is a reciprocal lattice vector (Squires 1978, Lovesey 1984). Measurement of the intensity of the Bragg scattering gives $(S_z)^2$, which is the square of the order parameter.

It is important to realize the difference between the elastic scattering as described in this section and the differential scattering as described in the previous section. In the static approximation, the differential scattering gives spin correlations at the time $\tau = 0$ (cf. Eq. 11.16), while the elastic scattering gives correlation over infinite times. The differential scattering gives both delta-function Bragg scattering from the long-range order at magnetic reciprocal lattice points and *diffuse* scattering (that is scattering diffuse in reciprocal space) from the short-range order. The elastic scattering consists of only the Bragg scattering part of the differential scattering.

Suggested Further Reading

Squires (1978)
Lovesey (1984)

12

BRAGG SCATTERING

12.1. The Scattering Geometry

In the previous chapter we saw that the magnetic scattering of neutrons from the ordered state included an elastic scattering term arising from correlations extending over infinite times and distances. In the absence of an applied magnetic field, such correlations are zero for the paramagnetic state and there is no elastic scattering term in the cross section. An analogous situation arises in the nuclear scattering; in the crystalline state there will be correlations in atomic positions extending over infinite times and distances; in the glassy state there will be correlations over infinite times but not over infinite distances; and for the liquid or gaseous state there will be zero correlation between atomic positions over infinite times or over infinite distances.

The correlation over infinite times give rise to a term $\delta(\omega)$ in the cross section, corresponding to elastic scattering. This delta function occurs in the scattering from the glassy state (and from spin glasses) as well as from the ordered magnetic state and from the crystalline state.

Correlation over infinite distances arise only in the crystalline state, where sums over terms involving $\exp(i\boldsymbol{\kappa} \cdot \mathbf{R})$ give rise to a delta function $\delta(\boldsymbol{\kappa} - \mathbf{g})$ in the scattering. This delta function arises also in the scattering of x-rays from crystals. Scattering from correlations over both infinite distances and infinite times is known as *Bragg scattering*. It involves a partial differential cross section that contains the terms

$$\delta(\omega) \sum_{\mathbf{g}} \delta(\boldsymbol{\kappa} - \mathbf{g}) \qquad (12.1)$$

so that the scattering is zero unless

$$\omega = 0 \qquad (12.2)$$

and

$$\boldsymbol{\kappa} = \mathbf{g} \qquad (12.3)$$

The definition of ω as given in Eq. 11.3 can be combined with Eq. 12.2 to show that

$$E' = E$$

and that

$$|\mathbf{k}| = |\mathbf{k}'| \qquad (12.4)$$

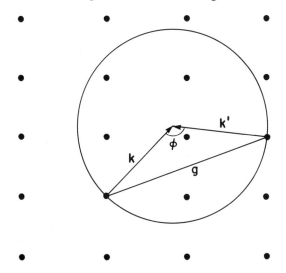

Figure 12.1. The geometry of Bragg scattering. For scattering to occur, the incident neutron wavevector **k**, the scattered neuron wavevector **k'**, and a reciprocal lattice vector **g** must form a triangle with two equal sides ($|\mathbf{k}| = |\mathbf{k'}|$). Bragg scattering occurs for a fixed incident wavevector **k** only when a reciprocal lattice point lies on the circle as shown.

The combination of Eq. 10.5 and Eq. 12.3 shows that

$$\boldsymbol{\kappa} = \mathbf{k} - \mathbf{k'} = \mathbf{g} \tag{12.5}$$

This equation is shown geometrically in Figure 12.1. For Bragg scattering to occur, **k**, **k'**, and **g** must form a triangle in reciprocal space. If ϕ is the angle through which the neutron is scattered, solution of the isosceles triangle in Figure 12.1 gives

$$g = 2k \sin(\phi/2) \tag{12.6}$$

This is Bragg's law. The law is usually expressed in a different form appropriate to real space. In real space we let λ be the neutron wavelength and d be the spacing of planes in the crystal, such that

$$k = \frac{2\pi}{\lambda}, \quad g = \frac{2\pi}{d} \tag{12.7}$$

In addition, we define an angle θ, known as the *Bragg angle* such that

$$\theta = \phi/2 \tag{12.8}$$

Then Eq. 12.6 becomes

$$\lambda = 2d \sin \theta \tag{12.9}$$

which is the form of Bragg's Law usually quoted.

If a neutron of wavevector **k** is incident on a crystal, we can determine whether Bragg scattering will occur by constructing a circle (or a sphere, in three dimensions) of radius k about the point $\mathbf{k}$ in reciprocal space. This is illustrated in Figure 12.1; if a reciprocal lattice vector is on the surface of the sphere, Bragg scattering will occur. The sphere is known as the *Ewald sphere*.

12.2. The Scattering Intensity

The treatment of the previous two chapters can be developed in a straightforward manner to derive the differential cross section for Bragg scattering from a crystalline sample (Squares 1978, Lovesey 1984). The magnetic elastic scattering for spin-only magnetism is given by

$$\frac{d\sigma}{d\Omega} = (\gamma r_0)^2 \frac{N(2\pi)^3}{v_0} \sum_{\mathbf{g}} \delta(\mathbf{\kappa} - \mathbf{g})$$
$$\times \sum_{\alpha\beta} (1 - \hat{\mathbf{\kappa}}_\alpha \hat{\mathbf{\kappa}}_\beta) F_\alpha^*(\mathbf{\kappa}) F_\beta(\mathbf{\kappa}) \qquad (12.10)$$

where $F_\alpha(\mathbf{\kappa})$ is known as the *magnetic structure factor*. It is given by

$$F_\alpha(\mathbf{\kappa}) = \sum_j S_{j\alpha} f_j(\mathbf{\kappa}) \exp(i\mathbf{\kappa} \cdot \mathbf{r}_j) \exp(-W_j) \qquad (12.11)$$

The sum over j is to be taken over all the magnetic atoms (at positions $\mathbf{r}_j$) within the magnetic unit cell. All the other terms are as defined in Chapters 10 and 11 with the addition of a term $\exp(-W_j)$, which is the Debye–Waller factor. This takes into account the fact that the jth atom is not fixed at its lattice site but vibrates about that site with amplitude $\mathbf{u}_j$. In fact

$$2W_j = \langle (\mathbf{\kappa} \cdot \mathbf{u}_j)^2 \rangle \qquad (12.12)$$

In like manner, the differential cross section for nuclear Bragg scattering can be calculated as

$$\frac{d\sigma}{d\Omega} = \frac{N(2\pi)^3}{v_0} \sum_{\mathbf{g}} \delta(\mathbf{\kappa} - \mathbf{g}) |F(\mathbf{\kappa})|^2 \qquad (12.13)$$

where $F(\mathbf{\kappa})$ is known as the *nuclear structure factor*. It is given by

$$F(\mathbf{\kappa}) = \sum_j \bar{b}_j \exp(i\mathbf{\kappa} \cdot \mathbf{r}_j) \exp(-W_j) \qquad (12.14)$$

with the sum over j to be taken over all the atoms (at positions $\mathbf{r}_j$) within the unit cell. By the measurement of nuclear and magnetic structure factors, we can attempt to determine nuclear and magnetic structures. This subject, however, is beyond the scope of this book. We nonetheless discuss two methods of actually measuring the structure factor, because these bring out points that are important for our narrative.

Crystal Rotation Method

Suppose a monochromatic neutron beam of wavelength λ and flux I_0 is incident on a small single crystal ("small" in this context means a crystal of area that is small compared with the area of the beam and a crystal that is sufficiently small that the beam is not significantly attenuated by the crystal). A counter is set to measure neutrons scattered through an angle ϕ, with ϕ determined by the Bragg condition (Eq. 12.6) for the reflection from the reciprocal lattice point **g**. The counter has an area that is sufficiently large that all neutrons Bragg-scattered from **g** enter it.

The crystal is rotated with uniform angular velocity through the position for Bragg scattering from **g** with the counter kept fixed. The integrated number of neutrons per unit time scattered into the counter, P, is then given by (Bacon 1975, Squires 1978, Lovesey 1984)

$$P = I_0 \frac{N}{v_0} \frac{\lambda^3}{\sin \theta} |F(\mathbf{g})|^2 \qquad (12.15)$$

for nuclear scattering. For magnetic scattering the term $|F(\mathbf{g})|^2$ has to be replaced by its magnetic analog, which can be written down readily by comparing Eq. 12.10 for the magnetic scattering with Eq. 12.13 for the nuclear scattering.

Laue Method

This technique involves the Bragg scattering of a "white" neutron beam from a small, fixed single crystal. Suppose the flux with wavelength between λ and $\lambda + d\lambda$ is $I_0(\lambda) \, d\lambda$; then the number of neutrons per unit time, P, entering a large counter appropriately positioned to collect all neutrons Bragg scattered from **g** is given by (Squires 1978, Lovesey 1984)

$$P = I_0 \frac{N}{v_0} \frac{\lambda^4}{2 \sin^2 \phi} |F(\mathbf{g})|^2 \qquad (12.16)$$

The values of λ and ϕ are those appropriate for Bragg scattering from the reciprocal lattice vector **g** in the particular orientation in which the crystal is fixed. Essentially, to satisfy the geometry of Bragg scattering the crystal can scatter only one particular wavelength λ in one particular direction for any given reflection **g**.

12.3. Monochromators and Analyzers

The Laue method is used to produce monoenergetic neutron beams from the white beams that emerge out of nuclear reactors. A single crystal, known as a *monochromator*, is placed in the beam out of a reactor in a suitable orientation to Bragg-scatter those neutrons of the required energy E. A suitable monochromator should be of a material from which large single crystals can be made; it should also have only very small absorption and incoherent cross sections so that the scattering is dominated by coherent scattering processes. Materials commonly used include beryllium, silicon, germanium, pyrolitic graphite, and copper.

A problem with crystal monochromators is that if a Bragg peak **g** is used, then $2\mathbf{g}, 3\mathbf{g}, \ldots, n\mathbf{g}$ form a set of Bragg peaks that satisfy the scattering geometry for incident wavevectors $2\mathbf{k}, 3\mathbf{k}, \ldots, n\mathbf{k}$ and for scattered neutron wavevectors $2\mathbf{k}', 3\mathbf{k}', \ldots, n\mathbf{k}'$, so that the "monochromatic" beam contains neutrons of all these wavevectors, that is neutrons of energy $E, 4E, 9E, \ldots, n^2E$. There will be very few neutrons for large n because the intensity varies as $I_0(\lambda) \sim \lambda^4$ (Eq. 12.16); λ^4 varies as n^{-4} and $I(\lambda)$ becomes small when λ is much less than the value at the peak of the Maxwelliam (cf. Figure 10.1). However, the presence in the beam of a component at $4E$ and sometimes also of a component of $9E$ does have to be allowed for. One method of eliminating the $4E$ component is to use reflections such as 111 or 311 from silicon or germanium, because in these cases the nuclear structure factor for 222 or 622 is zero and there is no Bragg peak for wavevector $2\mathbf{k}$.

Another technique for eliminating neutrons of energy n^2E from a "monochromatic" beam of energy E involves the use of neutron *filters*. The two most commonly used are polycrystalline beryllium cooled by liquid nitrogen (Iyengar 1965) and pyrolytic graphite (Loopstra 1966, Shirane and Minkiewicz 1970). The beryllium filter is almost transparent to neutrons with energy less than 5.2 meV and is opaque to neutrons of higher energy. It is used widely in connection with cold neutron beams. Pyrolytic graphite filters are opaque except for "windows" of transmission at around 9 meV and 13 meV.

Laue scattering is also used to measure the spectrum of the scattered neutron beam. A single crystal is placed in the scattered beam with the correct orientation to Bragg-scatter neutrons of some chosen wavevector and a counter is placed at the appropriate orientation to receive these scattered neutrons. By repeating this process for a series of wavevectors the spectrum of the scattered beam can be measured; such a process is referred to as analyzing the scattered beam, and the crystal involved is called an *analyzer*. Very much the same criteria apply in choosing an analyzer as in choosing a monochromator, since similar Laue-scattering processes are involved in each case.

12.4. Effects of Beam Collimation and of Mosaic Spread

In a real situation the neutrons in a beam do not all travel in exactly the same direction. The angular spread of the beam is called the *collimation*. In this section we examine the effect of collimation and of mosaic spread on the neutron beam.

Let us start with the effects of beam collimation. Suppose a neutron in the center of a beam with wavevector $\bar{\mathbf{k}}$ is Bragg-scattered through angle $\bar{\phi}$ by a crystal scattering from a reciprocal lattice vector of magnitude g. Then Eq. 12.6 shows that

$$g = 2\bar{k} \sin(\bar{\phi}/2) \qquad (12.17)$$

Now consider what would happen is another neutron were incident on the crystal traveling at a small angle $\Delta\phi$ to the right of the center line of

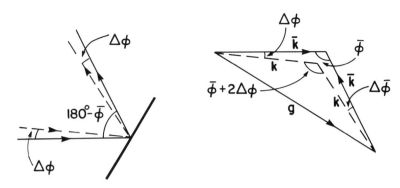

Figure 12.2. The effect of finite collimation on Bragg scattering. On the left, the scattering is shown in real space for a neutron traveling along the beam center line (solid line) and for a neutron traveling at an angle $\Delta\phi$ to this direction (dashed line). On the right, the same events are shown in reciprocal space.

the beam. The geometry of this is illustrated in real space on the left side of Figure 12.2 and in reciprocal space on the right side of the figure. Because the vector **g** is fixed in reciprocal space and the vector triangle is isosceles, the scattering angle must be $\bar{\phi} + 2\Delta\phi$. Bragg scattering will occur if

$$g = 2k \sin((\bar{\phi} + 2\Delta\phi)/2) \qquad (12.18)$$

The neutron that was initially traveling to the right of the center line of the beam, emerges to the left of the center line of the scattered beam.

Combining Eqs. 12.17 and 12.18 gives

$$k = \bar{k}\frac{\sin(\bar{\phi}/2)}{\sin(\bar{\phi}/2 + \Delta\phi)} \qquad (12.19)$$

We now consider the effect of crystal mosaic on beam collimation. Almost all real crystals are found to possess the property known as *mosaic structure*: this arises because the crystal lattice does not extend perfectly over macroscopic distances; rather the crystal consists of small perfect "mosaic" blocks with size of the order of a micrometer. These blocks are separated by dislocations, or rows of dislocations and in consequence are disoriented with respect to one another. Typical so-called "single-crystals" are found to possess a distribution of orientations of the lattice over angular spans that are in the range one minute to a few degrees. This is known as *mosaic structure*.

Let us consider what would happen if a neutron travelling along the center line of the beam is Bragg-scattered by a mosaic block at an angle $\Delta\theta$ to the central mosaic block. The geometry of the scattering process is illustrated in Figure 12.3. On the left is the picture in real space and on

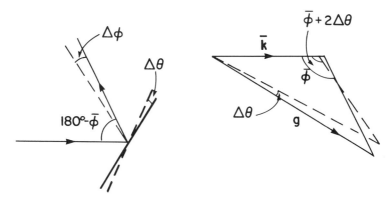

Figure 12.3. The effect of crystal mosaic on Bragg scattering. On the left the scattering of a neutron by a central mosaic block (solid line) and by a mosaic block at angle $\Delta\phi$ to the central block (dashed line) is shown in real space. On the right the same events are shown in reciprocal space.

the right is the corresponding picture in reciprocal space. The reciprocal-space diagram for the mosaic block consists of an isosceles triangle with sides k, k, and g and with scattering angle ϕ given by

$$\phi = \bar{\phi} + \Delta\phi = \bar{\phi} + 2\,\Delta\theta \tag{12.20}$$

Bragg's Law (Eq. 12.6) shows that

$$g = 2k\,\sin(\phi/2) \tag{12.21}$$

Combining Eq. 12.17, 12.20, and 12.21 gives

$$k = \bar{k}\,\frac{\sin(\bar{\phi}/2)}{\sin(\bar{\phi}/2 + \Delta\phi/2)} \tag{12.22}$$

Both Eq. 12.19 for collimation effects and Eq. 12.22 for mosaic effects give scattered neutron wavevectors whose magnitude is correlated with the angular direction of the scattered neutron. Figure 12.4 illustrates this for nominal scattering angles of 30 degrees and 90 degrees at a mean neutron wavelength of 2 Å. The upper panel shows the effects of collimation of the incident neutrons and the lower panel shows the effects of crystal mosaic. The energy spread in a beam of given angular width is greater for smaller scattering angles. This might, at first sight, suggest that monochromators and analyzers should always be used with large scattering angles to give more monochromatic scattering; however, matters are actually not so simple because the Laue cross section (Eq. 12.16) varies as $\sin^{-2}\phi$ so that large scattering angles lead to low beam intensities. In practice, as usual in experimental physics, some sort of compromise has to be made between beam intensity and resolution.

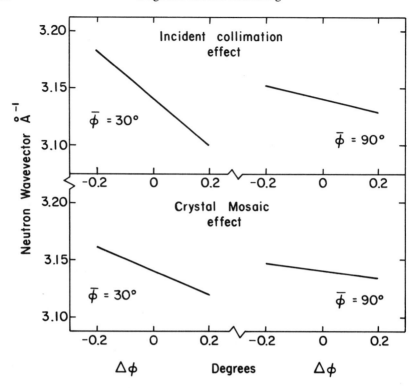

Figure 12.4. The correlation of the magnitude of the scattered neutron wavevector with the angle of scatter ϕ for neutrons of wavelength about 2 Å with scattering angles close to 30 degrees and to 90 degrees. The upper panel shows the correlation produced by collimation effects and the lower panel shows the correlation produced by crystal mosaic.

In actual neutron instruments involving crystal monochromators and analysers, there is a contribution to the resolution from both collimation and mosaic effects and there ceases to be a perfect correlation between neutron energy and $\Delta\phi$. However, it is apparent that a combination of the two effects will still give rise to strong correlations, since both involve correlations with $dk/d\phi$ negative (cf. Figure 12.4).

Resolution effects are particularly important in the measurement of critical scattering because the cross section varies very rapidly with κ near the critical point, and the quality of the resolution in reciprocal space will determine how closely it is possible to approach the critical point and still get meaningful data. The full treatment of resolution effects involves averaging the cross section over the resolution function in the four-dimensional space of κ and ω (Cooper and Nathans 1967, Nielsen and Moller 1969) and is only feasible with the aid of a computer. However, it should be pointed out that the correlation effects produced by Bragg scattering can be used to give *focusing effects,* whereby the broadening of sharp features in the cross section by resolution effects is minimized.

Suggested Further Reading

Als-Nielsen (1976a)
Bacon (1975)
Brown (1979)
Dachs (1978)
Iyengar (1965)

13

MEASUREMENT OF CRITICAL DYNAMICS

13.1. The Triple-axis Spectrometer

The triple-axis neutron spectrometer is shown in Figure 13.1. A "white" neutron beam is incident on a single-crystal monochromator that is aligned so as to reflect neutrons of wavevector k onto the sample. Neutrons scattered from the sample at an angle ϕ are incident on a single-crystal analyzer aligned so as to reflect neutrons of wavevector k' into a detector. The instrument is known as a triple-axis spectrometer because, before detection, a neutron is scattered about three parallel axes through angles $2\theta_M$, ϕ, and $2\theta_A$. These angles can all be varied continuously, as can the orientation ψ of the sample with respect to the direction of the incident neutron beam.

Except for the easily handled factor of k'/k, the scattering cross section is a function of κ and ω only (Eq. 11.10). This involves three variables (two components of κ and ω) so that only three of the four angular parameters need to be changed to set the instrument to detect the scattering for any required value of κ and ω. Usually either the scattering angle at the monochromator or at the analyzer are kept fixed while the other three angles are varied.

Collimators may be placed in any or all of the four neutron flight paths so as to better define the direction of the beams. Also, filters may be added if necessary to deal with such problems as n^2E components in the beam.

Because of the freedom to determine the scattering function at (almost) any desired value of κ and ω, we have the power to make measurements in whatever region is of physical interest. In magnetic critical scattering the interest usually lies in values of κ near to magnetic reciprocal lattice points, and measurements are taken by scanning the frequency ω at constant κ (Brockhouse 1961; sometimes referred to in the literature as *constant-Q scans*). Such scans constitute the normal way of taking data with the triple-axis spectrometer except in the special situation when the frequency response varies very rapidly as a function of κ; then the practice is to fix ω and to scan κ along some convenient linear path in reciprocal space. This is known as a *constant-energy scan*.

13.2. The Neutron Spin-echo Technique

The neutron spin-echo technique involves the use of polarized neutron beams, a topic that we have not so far introduced: indeed, we will not

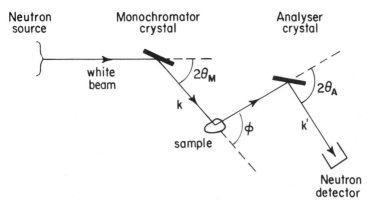

Figure 13.1. The triple-axis spectrometer. A monchromator crystal scatters just those neutrons from a white source with wavevector **k**. An analyzer crystal sends to the detector just those neutrons scattered through an angle ϕ that have wavevector **k'**.

attempt to describe the experimental techniques for producing and measuring polarized neutron beams. Suffice it to say that we produce a neutron beam polarized along the z direction and flip the polarization into the x direction at some point A along the beam. If the magnetic field along z is H_0, the polarization will undergo a Larmor precession in the x–y plane through angle ϕ. Over time τ a neutron of speed v will travel a distance l and the precession angle will be given by

$$\phi = \frac{\gamma_L l H_0}{v} \tag{13.1}$$

where $\gamma_L = 29.2$ MHz/T is the Larmor frequency (Mezei 1980a). This precession rate is large with typical applied fields and flight paths. For example, for neutrons of wavelength 4 Å in a field of 10^{-2} Tesla, the precession angle is 1830 radians per meter. If the neutron beam has a spread in velocity, the phase angle of the spin at any particular point on the flight path from A will soon become random; this is illustrated in Figure 13.2 taken from Mezei (1980b), who introduced this technique, and whose account we are following in this section. The figure shows how the spread in neutron speed in the beam leads to the average beam polarization in the x direction rapidly decaying to zero.

Although the beam becomes unpolarized, each individual neutron spin has been through a well-defined precession angle, as given by Eq. 13.1. The idea in the spin-echo technique is to suddenly reverse the direction of the field after neutrons have traveled a distance l_0. Then over a flight path of l_1 in a reversed field of H_1 the neutron spin will precess in the opposite direction. At point C where

$$H_0 l_0 = H_1 l_1 \tag{13.2}$$

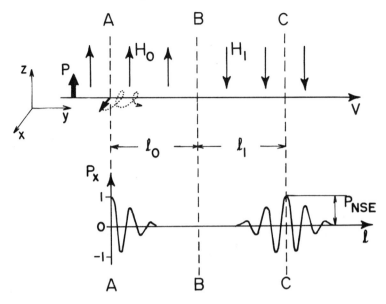

Figure 13.2. Larmor spin precession of neutrons in a beam and the spin-echo effect (Mezei 1980b).

the precession angle for each neutron in the beam will be the same and the beam will be fully polarized along the x direction again. This return of the polarization is known as the *spin echo*. Its existence has been shown experimentally (Mezei 1980b); in the experiments, one varies H_1 or H_0 keeping the flight paths fixed and one looks for the spin-echo point.

This technique can be applied also if the neutron is scattered at B and changes its velocity. So long as the scattered neutron's velocity in any particular direction is correlated with the incident neutron's velocity, there is a capability of observing the spin-echo (Mezei 1980b). The spin-echo gives a measure of a function $I(\kappa, \tau)$ that is the Fourier transform of the neutron-scattering function

$$I(\kappa, \tau) = \hbar \int S(\kappa, \omega) \cos[(\omega - \omega_0)\tau] \, d\omega \qquad (13.3)$$

where $\hbar\omega_0$ is the mean energy transfer on scattering: τ is a parameter with the dimensions of time; if the scattering function varies only weakly with κ and the energy transfer $\hbar\omega_0$ is small compared with the kinetic energy of the neutron E, then

$$\tau = \frac{\hbar\phi}{2E} \qquad (13.4)$$

We may vary τ by changing the precession fields. Typical values might be in the range 10^{-9} to 10^{-11} seconds, so that the spin-echo technique gives correlation functions in reciprocal space over this range of real times. Triple-axis spectrometry gives correlation functions in reciprocal space over shorter time scales, in the region 10^{-11} to 10^{-13} seconds (though the measurements are in terms of frequency rather than time). In practice, the two techniques are complementary and allow time (or frequency) scales to cover four order of magnitude (Mezei 1984).

Suggested Further Reading

Als-Nielsen (1976a)
Mezei (1980b)

14

TWO- AND ONE-DIMENSIONAL SYSTEMS

14.1. Introduction

The initial discovery of critical neutron scattering was made independently by Hughes and Palevsky (1953) and by Squires (1954). Both showed that the total cross section of iron increases markedly in the region of the critical temperature. Van Hove (1954a,b) suggested that this scattering was analogous to the critical scattering of light associated with the presence of large ordered regions and he developed a theory of the scattering in terms of a Ginzburg–Landau theory and of the correlation-function formalism developed by Ornstein and Zernicke (1914).

The remainder of this book is a review of what has been learned about critical phase transitions by the use of neutron scattering. The number of published measurements of critical neutron scattering is formidably large and no attempt is made to refer to them all: rather, a personal selection is made of the more significant works, as viewed from a present-day standpoint. The account is ordered according to the type of magnetic system involved, progressing from lower to higher dimensionality and from simpler to more complex Hamiltonians.

This chapter discusses magnetism in two- and one-dimensional systems. No one-dimensional system with finite-ranged interactions can support long-range order. However, the correlation length tends to infinity as the temperature tends to zero, as though there is a critical point at zero temperature and critical scattering is observed from the correlated regions.

In two dimensions there are phase transitions for spin-dimensionality 1 and 2. For the Ising model, there is an exact solution (Onsager 1944), while for the X–Y model the spins form vortices at low temperatures and there is a phase transition at which the vortices become bound (Kosterlitz and Thouless 1973, Kosterlitz 1974). For the X–Y model some physical properties show critical power-law behavior at the phase transition (e.g., δ and η) and others do not (e.g., ν and γ).

The one- and two-dimensional magnetic systems that are used are actually imbedded in real three-dimensional systems. The most-studied form of one-dimensional system arises from crystals of type ABX_3 where A is a nonmagnetic cation of single charge, B is a magnetic transition-metal cation of charge 2, and X is a halide anion. There is a simple hexagonal lattice with the transition-metal ions forming chains along the crystallographic c axis. Examples include $CsNiF_3$, $CsMnBr_3$, $CsCoBr_3$,

KCuF$_3$, and (CD$_3$)$_4$NMnCl$_3$ (TMMC). For the first four of these cases, the interactions along the chain are two to three orders of magnitude stronger than the interactions in the second and third dimensions, while for TMMC the large cation (CD$_3$)$_4$N$^+$ increases the separation between the chains, and the interactions in the second and third dimensions are four orders of magnitude smaller. All these materials give phase transitions to three-dimensional ordered states at sufficiently low temperatures, but there is a wide region of temperature above the crossover to the three-dimensional region, where the magnetic behavior is that of a one-dimensional system.

The type of two-dimensional magnetic system most studied by neutron scattering has chemical formula A$_2$BX$_4$ where, as with the one-dimensional systems, A is a cation with single charge, B is a magnetic transition-metal cation of charge 2, and X is a halide anion. The crystal structure is tetragonal with the magnetic ions forming a simple square lattice in two dimensions. The interactions in the third dimension are three to five orders of magnitude smaller than in the two-dimensional lattice. Examples of such materials include K$_2$NiF$_4$, K$_2$MnF$_4$, K$_2$CuF$_4$, K$_2$CoF$_4$, Rb$_2$CoF$_4$, and Rb$_2$CrCl$_4$.

The scattering from one-dimensional materials will show variations in only one dimension in reciprocal space (conventionally the z direction) and the correlation function $\hat{C}^{\alpha\beta}(\mathbf{q}, t, h)$ will depend on only $\mathbf{q}_z$ and will be independent of $\mathbf{q}_x$ and $\mathbf{q}_y$. In two-dimensional materials, the scattering shows variations in two dimensions in reciprocal space (conventionally the x–y plane) and the correlation function is independent of $\mathbf{q}_z$. The existence of these "rod" and "sheet" like properties of the critical correlations is fairly easy to check experimentally. The experiments confirm the reduced dimensionality, as expected from the nature of the Hamiltonian. In these circumstances it is possible to measure the static correlation function particularly accurately, since by arranging the scattering geometry such that k' is along the rod, or in the plane of the sheet, the scattering function $S^{\alpha\beta}(\mathbf{Q}, \omega)$ (Eq. 11.18) will be the same for all scattered neutrons and the static approximation becomes much better.

14.2. Two-dimensional Ising Systems

There is no two-dimensional magnet with Hamiltonian exactly corresponding to the Ising model. However, two materials, K$_2$CoF$_4$ and Rb$_2$CoF$_4$, have Hamiltonians that contain both Ising and Heisenberg antiferromagnetic terms (cf. Eqs. 4.1 and 4.3). The ratios of the Ising to the Heisenberg exchange parameter are 2.3 and 0.8, respectively (Breed 1969, Ikeda and Hutchings 1978), so that we might expect a crossover to occur in critical properties, with Ising-like behavior close to the critical point.

K$_2$CoF$_4$ and Rb$_2$CoF$_4$ show critical phase transitions at 107.8 K and at 102.6 K, respectively. The intensity of the magnetic elastic scattering (Eq.

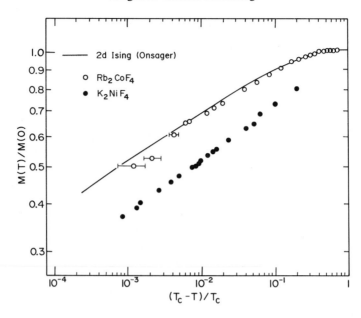

Figure 14.1. The reduced sublattice magnetization $M(T)/M(O)$ for K_2NiF_4 (closed circles) and for Rb_2CoF_4 (open circles) from Samuelsen (1973). The full line is the exact solution for the two-dimensional Ising model.

12.10) at the antiferromagnetic Bragg peak is directly proportional to the square of the staggered magnetization. Figure 14.1 shows results from the measurement of this intensity as a function of temperaure in Rb_2CoF_4; the staggered magnetization is plotted logarithmically against the reduced temperature and the results are compared with the predictions of Onsager's exact solution (Samuelsen 1973). The fit is excellent. Even at a reduced temperature of 10^{-3}, the magnetization is as much as one-half of its value at zero temperature. This is a consequence of the low value of the critical exponent β of the magnetization (one-eighth) for the two-dimensional Ising model. The same behavior of the magnetization with temperature is found in K_2CoF_4 (Ikeda and Hirakawa 1974).

Data such as that shown in Figure 14.1 can be used to determine the critical exponent β of the magnetization as a function of reduced temperature. For Rb_2CoF_4, independent measurements by Samuelsen (1973), Ikeda et al. (1979) and by Hagen and Paul (1984) gave values for β of 0.119 ± 0.008, 0.115 ± 0.016, and 0.114 ± 0.004, respectively, while for K_2CoF_4 Ikeda and Hirakawa (1974) find $\beta = 0.123 \pm 0.008$. The consensus of these measurements is that β is slightly lower than the exact value of 0.125 for the two-dimensional Ising model. In view of the excellent fit shown in Figure 14.1, and the similar fit of Ikeda and Hirakawa, it seems that the systems are behaving generally according to the two-dimensional Ising model, so that the discrepancy is hard to understand. One comment that should be made here is that it is hard to measure critical exponents to high accuracy by any method because the wide range of reduced temperature required imposes severe experimental difficulties and because the critical region may be limited in extent owing to the presence of correction terms (Eq. 3.3).

Extensive measurements have been made by neutron scattering of the static correlation function in K_2CoF_4 (Cowley et al. 1984). The static approximation should be excellent for Ising systems since, as pointed out in the introduction to Chapter 11, the time scale is infinitely long. The small Heisenberg term in the Hamiltonians will give a mechanism for dynamic behavior, but the time scale should be longer than for $X-Y$ or Heisenberg systems, so that the static approximation should still work well. This should enable good measurements to be taken of the correlation function in reciprocal space.

The exact solution of the two-dimensional Ising model enables the correlation function to be calculated explicitly (Wu 1966). At the critical temperature it varies asymptotically in real space as $R^{-1/4}$ for large distances so that $\eta = 1/4$ (cf. Eq. 5.26). Above the critical temperature the asymptotic behavior is as $R^{-1/2}\exp(-R/\xi)$ while below the critical temperature it is as $R^{-2}\exp(-R/\xi)$. Above the critical temperature this follows the Ornstein-Zernike form (Eq. 5.29) and after Fourier transformations to reciprocal space the correlations are reasonably approximated by the Lorentzian form of Eq. 5.35. At, and very near to, the critical temperature, the Fisher and Burford (1967) form of the correlation function (Eqs. 5.38 and 5.39) is a good approximation.

Below the critical temperature the form of the correlation function is different because the presence of long-range order modifies the nature of the decay of the correlation. Despite this asymmetry the critical exponent v of the correlation length ξ is symmetric above and below T_c. Fourier transformation of the real-space correlation function below T_c shows a form in which the Lorentzian is not as good as it is above T_c. Tarko and Fisher (1975) have given a better approximate form.

Figure 14.2 shows measurements by Cowley et al. (1984) of the correlation function in K_2CoF_4 in reciprocal space at temperatures of 108.08 K and 110.39 K ($T_c = 107.72$ K). Near the critical point the correlation length becomes large and the correlation function is sharply peaked in reciprocal space. In this case the effects of experimental resolution may become marked and this must be taken into account in interpreting the data. The lines in the figures are fits to the Lorentzian form convoluted with the resolution function. The fit is excellent.

Figure 14.3 shows a measurement by Cowley et al. (1984) of the correlation function in K_2CoF_4 in reciprocal space at a temperature of 107.30 K ($T_c = 107.72$ K). The broken line shows the fit to the Lorentzian form and the full line shows the fit to the Tarko-Fisher form, both convoluted with the experimental resolution function. The Lorentzian form does not give too bad a fit, but the Tarko-Fisher form is clearly better. Figure 14.3 also shows data at 105.37 K where again the Tarko-Fisher form is superior.

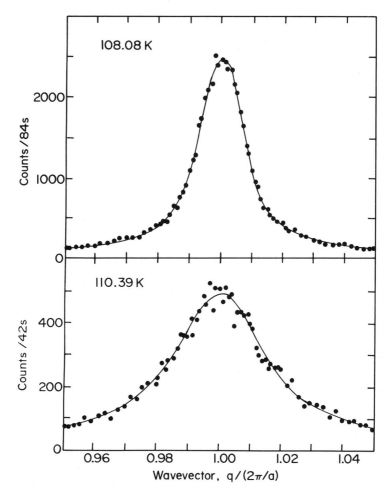

Figure 14.2. The critical scattering observed in K_2CoF_4 at two temperatures above T_c ($T_c = 107.72$ K) for $\kappa = (q, 0, 0.45)$. The lines are fits to the Lorentzian form convoluted with the resolution function (from Cowley et al. 1984).

All physically reasonable critical correlation functions decay monotonically in real space, and, after Fourier transformation, give a form in reciprocal space that decays monotonically from reciprocal lattice points. The scattering technique is not very sensitive in picking up the power of R in the correlations (Eq. 5.29) since, as we have seen, it can only just distinguish between $R^{-1/2}\exp(-R/\xi)$ and $R^{-2}\exp(-R/\xi)$. To go on to make a determination of the correlation length ξ and its critical exponent ν from the scattering depends on knowledge of the form of the correlations, and normally this form is not known, so an approximate form must be assumed. Cowley et al. show that fits for ν' using the Lorentzian form give $\nu' = 0.86 \pm 0.06$ with $\chi^2 = 1.10$, while for the

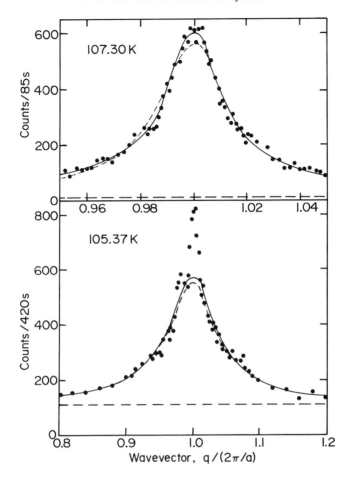

Figure 14.3. The critical scattering observed in K_2CoF_4 at two temperatures below T_c ($T_c = 107.72$ K) for $\kappa = (q, 0, 0.45)$. The horizontal broken line shows the background and the full and broken lines the result of a least-squares fit to the Tarko–Fisher and the Lorentzian forms, respectively, convoluted with the experimental resolution (from Cowley, Hagen and Belanger 1984).

Tarko–Fisher form $v' = 1.12 \pm 0.13$ with $\chi^2 = 0.40$. It is probably impossible to determine v' accurately in the (usual) case in which the form of the correlation function is unknown. Some doubts are also raised about the determination of the index v above T_c, but here the problem is less severe since the Lorentzian form seems to be a better approximation.

Table 14.1 shows the results of measurements of the critical exponents v, v', γ, γ', η, and β in Rb_2CoF_4 and K_2CoF_4. The exponent γ is determined from the amplitude of the correlation function at $\mathbf{q} = 0$ (cf. Eq. 5.37) and η is determined from the form of the correlation function at the critical temperature.

Cowley et al. (1984) and Hagen and Paul (1984) have gone on to look

Table 14.1. Critical Exponents of Two-dimensional Ising Systems as Determined by Neutron-scattering Measurements

Material	Exact Solution	Rb_2CoF_4	Rb_2CoF_4	Rb_2CoF_4	K_2CoF_4	K_2CoF_4
Reference	—	Samuelsen (1973)	Ikeda et al. (1979)	Hagen and Paul (1984)	Ikeda et al. (1974)	Cowley et al. (1984)
ν	1.00	0.89 ±0.1	0.99 ±0.04	—	0.97 ±0.04	1.02 ±0.05
ν'	1.00	—	—	—	—	1.12 ±0.13
γ	1.75	1.34 ±0.22	1.67 ±0.09	—	1.71 ±0.04	1.73 ±0.05
γ'	1.75	—	—	—	—	1.92 ±0.20
η	0.25	—	0.2 ±0.1	—	—	—
β	0.125	0.119 ±0.008	0.115 ±0.016	0.114 ±0.004	0.123 ±0.008	—

at the constant terms multiplying the power-law terms in the susceptibility, the correlation length, and the magnetization. Determination of the first two of these is sensitive to the analytic form of the correlation function, but when appropriate fitting are made there is satisfactory agreement with theory in all three cases.

The work with two-dimensional Ising systems is as much a proving ground for the neutron-scattering technique as it is a test of the two-dimensional Ising model. Experiments done by different groups show reasonable consistency and the sort of accuracy that can be obtained is apparent.

14.3. Two-dimensional $X-Y$ Systems

As with the Ising case, there is no two-dimensional magnet with Hamiltonian corresponding exactly to the $X-Y$ model. However, there are a number of materials with Hamiltonians that contain both $X-Y$ and Heisenberg terms and that cross over to $X-Y$-like behavior at low temperatures. The most studied of these is K_2CuF_4, which has a Hamiltonian whose largest term corresponds to a two-dimensional Heisenberg ferromagnet. The $X-Y$ behavior comes from a ferromagnetic term that is about 1% of this Heisenberg term, while three-dimensional interactions are another order of magnitude smaller (Hirakawa and Ikeda 1973, Moussa et al. 1978). These are conditions under which crossover behavior is to be expected, and indeed this is observed experimentally (Hirakawa 1982), with two-dimensional Heisenberg behavior above about 7.3 K, two-dimensional $X-Y$ behavior between about 7.3 K and 6.6 K, and three-dimensional $X-Y$ behavior below about 6.6 K (all in zero magnetic field). The critical temperature is 6.25 K and extrapolation in the two-dimensional X–Y region indicates that a two-dimensional phase transition would occur at about 5.5 K if there were no crossover to three-dimensional behavior. The critical behavior of a system with two crossover regions is clearly not simple and we concentrate on the temperature region between 6.6 K and 7.3 K where the properties should follow those of the two-dimensional X–Y model.

Kosterlitz (1974) has estimated the critical properties of the two-dimensional X–Y model. He finds that

$$\xi \sim \exp(bt^{-1/2}) \qquad t > 0 \qquad (14.1)$$

$$\chi \sim \exp(bt^{-1/2}(2-\eta)) \qquad t > 0 \qquad (14.2)$$

with

$$\eta = 1/4 \quad \text{and} \quad b \approx 1.5 \qquad (14.3)$$

where η is the critical exponent of the correlation function as defined by Eq. 5.27. The correlation length ξ and the susceptibility χ do not have the usual form (cf. Table 3.1), so that the exponents ν and γ are not defined. Figure 14.4 shows a plot of $\ln \chi$ versus $t^{-1/2}$ (Hirakawa 1982);

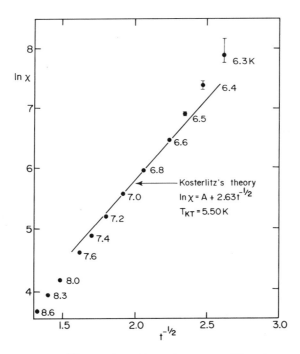

Figure 14.4. Ln χ versus $t^{-1/2}$ for K_2CuF_4 (from Hirakawa 1982). In the temperature range 6.6 K to 7.3 K the susceptibility follows the predictions of Kosterlitz and Thouless for a two-dimensional X–Y system. At higher temperatures there is a crossover to two-dimensional Heisenberg behavior and at lower temperatures there is a crossover to three-dimensional X–Y behavior.

according to Eq. 14.2 this should be a straight line of slope $b(2 - \eta) \approx 2.63$, and this is indeed found to be the case between 6.6 K and 7.3 K. The measurement of ξ and of η has been seen to depend, at least weakly, on the choice of analytic form for the correlation function. Hirakawa et al. (1982) took the form given by Eq. 5.38, and Figure 14.5 shows their plot of the correlation function $\hat{C}$ as a function of the in-plane component of q (denoted as q_a in the figure) at 6.93 K. The fit is best if $\eta = 0.25$ as predicted by Kosterlitz. Only in the temperature region appropriate for the two-dimensional $X-Y$ model is η found to have this value.

From these fits one can also find the correlation length ξ (or

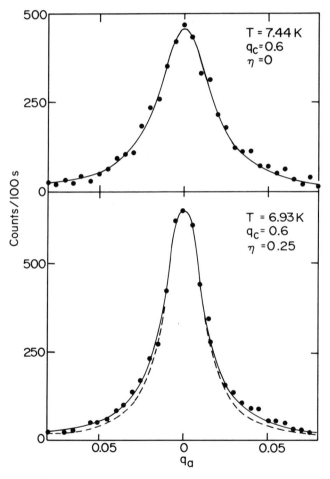

Figure 14.5. The critical scattering observed in K_2CuF_4 at two temperatures above T_c for $\kappa = (q_a, 0, q_c)$ from Hirakawa et al. (1982). The solid lines are fits to the Fisher–Burford form convoluted with the resolution function. At $T = 6.93$ K, in the two-dimensional $X-Y$ region, the fit gives $\eta = 0.25$ as predicted; at $T = 7.44$ K the system is in the region of crossover to Heisenberg behavior and the fit gives $\eta = 0$.

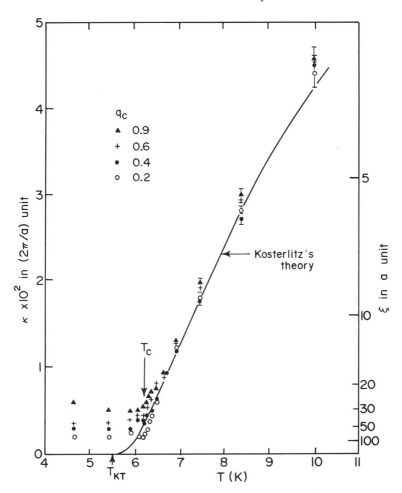

Figure 14.6. Inverse correlation length versus temperature in K_2CuF_4 (Hirakawa 1982) as derived from scans such as those shown in Figure 14.5. Below about 6.3 K the correlations depend on q_c so that the behavior is three-dimensional. At higher temperatures the correlations are two-dimensional and follow the predictions for the two-dimensional X–Y model.

alternatively $\kappa_1 = 1/\xi$) and compare with Kosterlitz's predictions; this is done in Figure 14.6 (Hirakawa et al. 1982). At lower temperatures the effect of three-dimensional correlation affects the fits as the scattering depends on q_c, the component of **q** out of the two-dimensional plane, but in the region of temperature from 6.6 K to 7.3 K the fit is excellent.

Although less extensively studied, materials of type $BaM_2(XO_4)_2$ with M = Co or Ni and X = P or As have more pronounced X–Y character while still being predominantly two-dimensional in character (Regnault et al. 1983). These materials have the honeycomb lattice in two dimensions and in, for example, $BaNi_2(PO_4)_2$ the X–Y term is 23% of the

Heisenberg term. The critical temperature is 23.5 K and the critical properties cross over to being two-dimensional at about 23.7 K ($t \sim 10^{-2}$). Extrapolation indicates that a two-dimensional phase transition would occur at 22.6 K ($0.96T_c$) were there no crossover to three-dimensional behavior. This gives a considerably wider region of two-dimensional X–Y behavior than is the case for K_2CuF_4; a complication is that the exchange interaction in the planar lattice is not just to nearest neighbors but extends significantly to at least third-nearest neighbors. This might be expected to narrow the critical region. Figure 14.7 shows the fitted plot of the correlation length to the Kosterlitz–Thouless exponential form of Eq. 14.1 with $b = 1.6$ and a transition temperature of $0.96T_c$. The fit is excellent.

A more rigorous test of the Kosterlitz–Thouless theory must await the discovery of a material whose magnetic Hamiltonian corresponds more

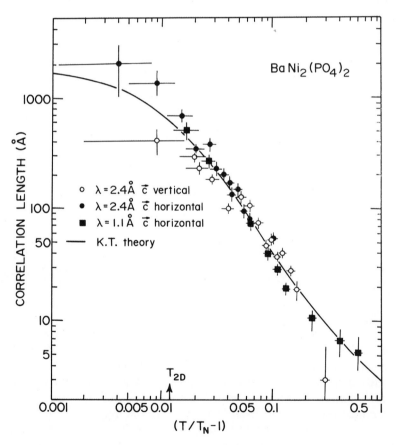

Figure 14.7. Temperature dependence of the in-plane correlation length in $BaNi_2(PO_4)_2$ from Regnault et al. (1983). There is an excellent fit to the predictions of the Kosterlitz–Thouless theory for the two-dimensional X–Y model (solid line).

closely to the two-dimensional $X-Y$ model. However, the works described above fit in well with the theory and tend to confirm it.

Moussa et al. (1978) and Hirakawa et al. (1983) have investigated the spin dynamics of K_2CuF_4. In two dimensions the short-range order is greater at and above T_c for fixed R or q than it is in three dimensions, and this makes it easier for the system to support spin-wave excitations at and just above the critical temperature.

For long wavelengths, the spin-wave energy drops rapidly just as the critical temperature is approached from below and the critical temperature is somewhere close to the point of critical damping. At shorter wavelengths, the spin-wave energies change only slightly on passing through T_c, but become damped out when the correlation length ξ becomes smaller in extent than about one wavelength. Because of the mixed nature of the Hamiltonian, it is likely that the dynamics (and statics) at T_c will follow a three-dimensional model at smaller wavevector $\mathbf{q}$, a two-dimensional $X-Y$ model at intermediate wavevectors, and a two-dimensional Heisenberg model at large wavevectors. The $X-Y$ region is too small for a full investigation to be made of $X-Y$ critical dynamics.

A similar persistence of short-wavelength spin-wave excitations above the critical temperature has been obseved by Hutchings et al. (1986) in another two-dimensional $X-Y$ material, Rb_2CrCl_4.

14.4. Almost Two-dimensional Heisenberg Systems

There are a number of two-dimensional magnets for which the Heisenberg term is much larger than the Ising term, and in which the three-dimensional interactions are extremely small. The two materials most extensively studied are K_2NiF_4 (Birgeneau et al. 1971a and references therein) and K_2MnF_4 (Birgeneau et al. 1973 and references therein). In these materials there is a single-ion anisotropy term in the Hamiltonian with a magnitude of 0.2% and 0.4%, respectively, relative to the Heisenberg term: this term acts to align the spins in the z direction, so the materials would be expected to cross over to Ising behavior close to the critical point.

In Figure 14.1 the magnetization of K_2NiF_4 was plotted as a function of the reduced temperature and compared with Rb_2CoF_4. The magnetization of K_2NiF_4 is always lower, but its critical exponent, as measured by the slope of the plot, is close to the Ising value and Birgeneau et al. (1971a) give $\beta = 0.138 \pm 0.004$. Figure 14.8 shows a similar plot for K_2MnF_4 (Birgeneau et al. 1973), this time extended over four orders of magnitude of reduced temperature, and the best fit gives $\beta = 0.15 \pm 0.01$. Similar experiments by Ikeda and Hirakawa (1972) yielded $\beta = 0.188 \pm 0.01$, though Birgeneau et al. (1973) suggest that the sample of Ikeda and Hirakawa shows some broadening of the critical point owing to chemical inhomogeneity. It is clear that the experimental results all indicate values

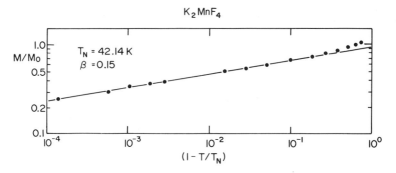

Figure 14.8. Reduced sublattice magnetization $M(T)/M(O)$ in K_2MnF_4 as a function of reduced temperature (from Birgeneau et al. 1973). The solid line is a best fit of the data to a power law with $T_N = 42.14$ K and $\beta = 0.15$.

of β that are higher than the value of 0.125 for two-dimensional Ising systems.

The critical scattering above T_c has been measured in K_2NiF_4 and K_2MnF_4 by Als-Nielsen et al. (1976a). The scattering is found to have two components, one from the Ising-like zz correlations and the other from the xx and yy correlations. Only the first of these components diverges at T_c. The measurements of the inverse zz correlation length are plotted logarithmically against reduced temperature in Figure 14.9. For both K_2NiF_4 and K_2MnF_4 there is power-law behavior with $\nu = 0.90 \pm 0.10$. This is, within experimental error, the same exponent as for the two-dimensional Ising model ($\nu = 1$). In similar manner, Als-Nielsen et al. found $\gamma = 1.65 \pm 0.15$ for both materials, again—within error—the same result as for the two-dimensional Ising model ($\gamma = 1.75$). However, as Figure 14.9 shows, the absolute value of the correlation length is not the same as is found in the two-dimensional Ising model.

An interesting aside to this matter is that Stanley and Kaplan (1966) showed that series expansions for the two-dimensional Heisenberg model indicate the presence of a critical phase transition, even though there is an exact proof that there can be no long-range ordered state (Mermin and Wagner 1966). Use of the experimentally determined Heisenberg exchange parameter for K_2MnF_4 and K_2NiF_4 yields predicted critical temperatures, from the equations of Stanley and Kaplan, of 41.8 K and 93.1 K, while the observed critical temperatures are 42.1 K and 97.1 K. The two-dimensional Heisenberg model is an unsolved problem and the question of the existence of a phase transition is still controversial (Guttmann 1978, Praveczki 1980, Shenker and Tobochnik 1980). It is not even clear how relevent this question is to K_2MnF_4 and K_2NiF_4, since the observed critical scattering comes primarily from $\hat{C}^{zz}(\mathbf{q}, t, h)$. However, the crossover to Ising-like behavior only takes place quite close to T_c and may only have a small influence on the value of T_c. The critical

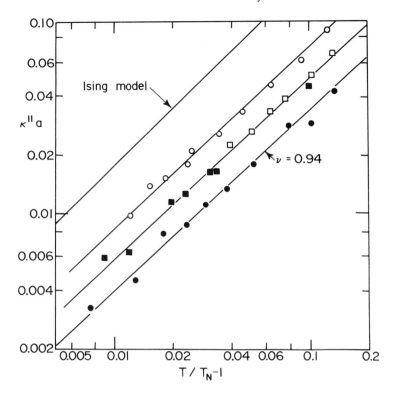

Figure 14.9. Logarithmic plot of the correlation lengths in $Rb_2Mn_{0.5}Ni_{0.5}F_4$ (open circles), K_2NiF_4 (filled circles) and K_2MnF_4 (open and filled squares), together with the exact result for the two-dimensional Ising model (from Als-Nielsen et al. 1976a).

temperatures predicted by Stanley and Kaplan seem too close to the observed temperatures for coincidental equalities to be invoked.

The critical dynamics of K_2NiF_4 and of K_2MnF_4 have been investigated by Birgeneau et al. (1971a and 1973, respectively). Since the out-of-plane (zz) component of the static correlations is very different from the in-plane (xx and yy) components, consideration of the dynamics must be split into two corresponding parts.

The in-plane correlations give rise to spin waves. At large wavevector **q** the spin waves show no explicit recognition of the critical temperature as the temperature is changed; that is, there is no sudden change of the energy, the line width, or the intensity. In K_2MnF_4 the spin-wave energy is about 9% lower at the critical temperature than at low temperatures. For higher temperatures, the spin-wave peaks broaden out and become overdamped (such as in K_2CuF_4). For spin waves of small wavevector below the critical temperature, there is an energy gap due to the anisotropic nature of the Hamiltonian. The limitingly small-wavelength spin waves remain sharp as the temperature varies below T_c with an energy that varies in the same way as the staggered magnetization.

The out-of-plane correlations give a dynamic response that is peaked at zero frequency. At and below T_c this peak has an apparent width equal to the experimental resolution function, but as the temperature rises above T_c the width increases.

14.5. One-dimensional Systems

Since one-dimensional systems cannot support long-range order, there is only a limited amount to be learned about critical properties from their

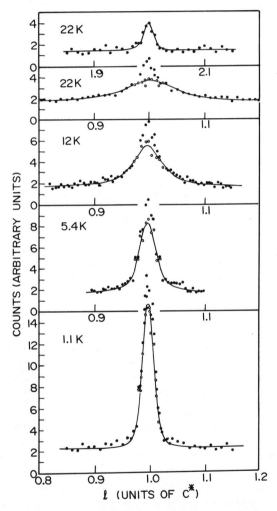

Figure 14.10. The critical scattering in the one-dimensional antiferromagnet TMMC at temperatures between 1.1 K and 22 K with $\kappa = (0.28, 0, 1)$ for the upper four scans and $\kappa = (0.1, 0.1, 1)$ for the lower scan (from Birgeneau et al. 1971b). The closed circles are actual experimental data; the open circles give the residual magnetic scattering after the nuclear contribution has been subtracted. The upper scan shows the nuclear scattering and the lower four scans show magnetic scattering. The lines are fits to Lorentzian correlation functions convoluted with the instrumental resolution function.

study. However, one-dimensional systems do exhibit strong correlations and large correlation lengths at low temperatures. These properties are characteristic of the critical region and it is as though there is an extended critical region with a critical point at zero temperature; in fact, the correlation length diverges in the limit as the temperature tends to zero.

Figure 14.10 shows the critical scattering from TMMC at temperatures varying from 1.1 K to 22 K (Birgeneau et al. 1971b). The magnetism is one-dimensional at all these temperatures and the solid lines are Lorentzian peaks (Eq. 5.35) convoluted with the instrumental resolution function. The Lorentzian form is satisfactory; it is clearly broadening and losing intensity as the temperature rises, showing weakening correlations and increasing Lorentzian width κ_1 ($=1/\xi$). These measurements agree well with the exact solution of the one-dimensional Heisenberg model for $S = \infty$ (Fisher 1964). Similar good fits to exact solutions have been obtained by Fitzgerald et al. (1982) for $CsMnBr_3$ (Heisenberg model), by Steiner and Dachs (1971) for $CsNiF_3$ (X–Y model), and by Kopinga et al. (1985) for $CsFeCl_3 \cdot 2D_2O$ (Ising model). We show just the latter of these results in Figure 14.11. The solid line is the prediction of the Ising model and the dashed line is the prediction of a model with a Hamiltonian given by a Heisenberg term, with $|J/k_B| = 6.0$ K, plus an anisotropic term $D \sum (S_i^z)^2$ with $D/k_B = -40$ K. The second of these is the better fit.

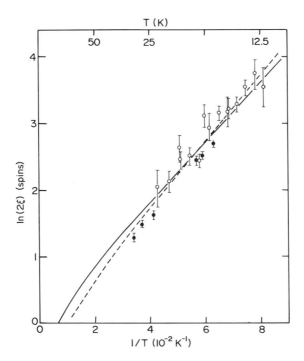

Figure 14.11. The logarithm of the correlation length ξ versus the reciprocal of the temperature for the one-dimensional antiferromagnet $CsFeCl_3 \cdot 2D_2O$ from Kopinga et al. (1985). There is an approximately straight-line variation showing an exponential growth of ξ with T^{-1}.

We can conclude that critical correlations in these one-dimensional systems are understood quantitatively as well as qualitatively. The correlations are sufficiently long ranged that one-dimensional systems exhibit well-defined spin-wave excitations for wavelengths less than the correlation range. These is no exact theory of the dynamics of one-dimensional systems but experiments seems to fit well to approximate spin-wave theories (Steiner et al. 1976, Gaulin and Collins 1984). A full treatment of the dynamics of one-dimensional systems involves the consideration of nonlinear soliton excitations and of a continuum of states under the spin-wave excitations, but coverage of this will not be attempted in this book.

Suggested Further Reading

Als-Nielsen (1976a)
de Jongh and Midiema (1974)
Steiner et al. (1976)

15

ISING SYSTEMS IN THREE DIMENSIONS

15.1. Magnetic Systems

As in two dimensions, there is difficulty finding a material with Hamiltonian corresponding to the Ising model. Good correspondence is found in the rare-earth materials dysoprosium aluminum garnet (DAG) and lithium terbium fluoride (LiTbF$_4$), where the crystal-field effect is such as to give ground states of $|J_z = \pm 15/2\rangle$ and $|J_z = \pm 6\rangle$, respectively for the rare-earth ions, with other states at energies more than $k_B T_c$ higher. This is isomorphous with the spin-half Ising system where $|S_z = \pm 1/2\rangle$, as the other states have negligible influence on the critical properties. The predominant interactions in both these materials are dipolar. In DAG these interactions give rise to an antiferromagnetic structure at low temperatures, but a complex antiferromagnet in which the local z direction is not the same at all dysprosium sites. LiTbF$_4$ is a ferromagnet at low temperatures.

Other magnetic materials that correspond to the Ising model are MnF$_2$ and FeF$_2$. These both have the rutile crystal structure, which gives a body-centered tetragonal lattice. There is antiferromagnetic Heisenberg exchange between the moments at the corner and at the body-centered sites that gives rise to an ordered antiferromagnetic lattice at low temperatures. In addition, there is an anisotropic term in the Hamiltonian that tends to align the moments along the tetragonal axis. In MnF$_2$ the anisotropy energy arises predominantly from dipolar interactions and is about 35 times smaller than the exchange energy. In FeF$_2$ the anisotropy energy arises predominantly from crystal-field effects that give a term $D \sum_i (S_i^z)^2$ in the Hamiltonian (with $D < 0$). The same dipolar effects are present in FeF$_2$ as in MnF$_2$, but the crystal field term is larger by an order of magnitude, and is in fact about 35% of the Heisenberg term. Thus, we might expect to see crossover effects in MnF$_2$, from Heisenberg behavior far from the critical point to Ising behavior near to the critical point. In FeF$_2$ we might expect the critical region to be almost entirely Ising-like.

In three dimensions, dipolar interactions (which vary as R^{-3}) cannot be regarded as short range, so their presence may change the universality classifications of a system. Aharony (1973a,b) has shown that this is indeed the case for ferromagnetic order, but for antiferromagnetic order the long-range nature of the interactions becomes irrelevant in the renormalization group recursion relations.

15.2. Ising Antiferromagnets

We first describe the three antiferromagnetic systems for which the universality classification does seem to correspond to the Ising model. In all three cases it is found that only the zz component of the correlation function $\hat{C}^{\alpha\beta}(\mathbf{q}, t, 0)$ diverges at the critical point, as would be expected for an Ising system. In DAG, scattering is only seen from the zz correlation (Norvell et al. 1969b), in FeF$_2$ transverse spin fluctuations can be detected but vary very little in the critical region (Hutchings et al. 1972). The fluctuations in MnF$_2$ show typical crossover properties with the correlation length behaving as shown in Figure 8.2. The shift in critical temperature caused by the anisotropy is about 1.3 K (Dietrich 1969), while the critical temperature is 67.5 K.

Belanger and Yoshizawa (1987) find for FeF$_2$ that the correlation function is reasonably fitted by a Lorentzian form (Eq. 5.35) except very close to the critical point. For reduced temperatures with magnitude less than 10^{-3}, the forms of Fisher and Burford (1967) (Eq. 5.38) and of Tarko and Fisher (1975) for temperatures above and below T_N, respectively, are found to give better fits if η is assumed to have a value appropriate for the Ising model in three dimensions. However, the differences were not large enough to enable a distinction to be made between various possible approximate forms (such as Eq. 5.38 with $\phi = 1$ or with ϕ given by Eq. 5.39, or Eq. 5.40).

Table 15.1 gives the measured critical exponents of these three antiferromagnets. The predictions of the three-dimensional Ising model are shown for comparison. The overall agreement is satisfactory. The most accurate measurements are for FeF$_2$ (Belanger and Yoshizawa 1987) and these fit the Ising model predictions very well. The fitting assumed $\gamma = \gamma'$ and $\nu = \nu'$ as predicted by scaling; the reason for making

Table 15.1. Measured Critical Exponents in the Antiferromagnets FeF$_2$, MnF$_2$, and DAG, Compared with the Predictions of the Ising Model in Three Dimensions (Table 5.1)

Exponent	FeF$_2$	MnF$_2$	DAG	Ising
β	0.325 ± 0.010[a]	0.335 ± 0.005[c]	0.26 ± 0.02[e]	0.326
γ	1.25 ± 0.08[b]	1.27 ± 0.02[d]	1.16 ± 0.06[f]	1.240
γ'	1.25 ± 0.02[b]	1.32 ± 0.06[d]	—	1.240
ν	0.64 ± 0.01[b]	0.63 ± 0.02[d]	0.61 ± 0.02[f]	0.631
ν'	0.64 ± 0.01[b]	0.56 ± 0.05[d]	—	0.631
η	—	0.05 ± 0.02[g]	—	0.039

[a] Wertheim (1967) (Mössbauer).
[b] Belanger and Yoshizawa (1987); fitted with $\nu' = \nu$ and $\gamma' = \gamma$.
[c] Heller and Benedek (1962); Heller (1966) (NMR).
[d] Schulhof et al. (1971).
[e] Norvell et al. (1969a).
[f] Norvell et al. (1969b).
[g] Schulhof et al. (1970).

this assumption is that a temperature-independent amplitude ratio for the susceptibility and correlation length above and below T_N can then be found. The experimental values for these ratios are in satisfactory agreement with the predictions of the Ising model.

By measuring the transverse correlation length and using the shifted critical temperature (cf. Section 8.1) Schulhof et al. (1970) find exponents $v = 0.63 \pm 0.08$ and $\gamma = 1.47 \pm 0.10$ for the region beyond the crossover in MnF_2. We would expect this region to be Heisenberg-like with $v = 0.71$ and $\gamma = 1.39$ (Table 5.1).

Overall, it must be concluded that the static critical properties of DAG, FeF_2, and MnF_2 are as expected theoretically.

We turn now to the dynamic critical properties of FeF_2 and MnF_2, and investigate the dynamic part of the correlation function, which is represented by the function $F(\mathbf{q}, t, \omega)$ defined in Eq. 7.3. It is apparent that this function will have two forms, $F_z(\mathbf{q}, t, \omega)$ for the zz (or longitudinal) correlations and $F_x(\mathbf{q}, t, \omega)$ for the xx and for the yy (or transverse) correlations. The mixed (Ising and Heisenberg) nature of the Hamiltonian has a major influence on the transverse correlations and leads, as we have shown, to two correlation lengths in the system, ξ_z and ξ_x, with reciprocals κ_z and κ_x respectively. This leads not only to the crossover in thermodynamic properties that we have seen, essentially corresponding to $\mathbf{q} \to 0$ in reciprocal space, but also to a crossover in critical properties right at T_c from "Ising" (or anisotropic) behavior to "Heisenberg" (or isotropic) behavior as the wavevector $\mathbf{q}$ passes from being smaller than κ_x to larger. In MnF_2 this feature makes the Ising critical region rather restricted for the measurement of dynamic properties, because experimental resolution functions are larger for dynamic than for static measurements. Only in FeF_2 has it proved possible to make extensive measurements in the Ising region and we describe this case primarily.

The longitudinal dynamic correlations show a peak at $\omega = 0$ for constant $\mathbf{q}$ and t known as a central mode. At small ω it is approximated for fitting purposes by the Lorentzian function:

$$F_z(\mathbf{q}, t, \omega) = \frac{\Gamma_z(\mathbf{q}, t)}{\Gamma_z(\mathbf{q}, t)^2 + \omega^2} \qquad (15.1)$$

The transverse correlations show spin-wave behavior at low temperatures. We denote the spin-wave energy by $\omega_0(\mathbf{q}, t)$. At high temperatures the form becomes the same as for the longitudinal correlations and we adopt the approximate expression

$$F_x(\mathbf{q}, t, \omega) = \frac{\Gamma_x(\mathbf{q}, t)}{\Gamma_x(\mathbf{q}, t)^2 + (\omega - \omega_0(\mathbf{q}, t))^2}$$
$$+ \frac{\Gamma_x(\mathbf{q}, t)}{\Gamma_x(\mathbf{q}, t)^2 + (\omega + \omega_0(\mathbf{q}, t))^2} \qquad (15.2)$$

where, as $t \to \infty$, $\Gamma_x(\mathbf{q}, t) \to \Gamma_z(\mathbf{q}, t)$ and $\omega_0(\mathbf{q}, t) \to 0$.

Riedel and Wegner (1970) have extended dynamic scaling to include mixed systems like FeF_2 and MnF_2. They identify the characteristic frequency ω_c for the longitudinal fluctuations as $\Gamma_z(\mathbf{q}, t)$ and write that, instead of being a function of just two variables, $\omega_c(\mathbf{q}, t)$ or alternately $\omega_c(\mathbf{q}, \kappa)$, the characteristic frequency now depends on three variables, $\omega_c(\mathbf{q}, \kappa_z, \kappa_x)$. No attempt will be made to develop this theory, but it leads to a scaling equation identical to Eq. 7.7 for the longitudinal characteristic frequency with $z = 2$. For the Heisenberg antiferromagnet, in contrast, $z = 1.5$ (Eq. 7.21). For $z = 2$, we get

$$\begin{aligned}\Gamma_z(q, t) &= \kappa_z^2 f(q/\kappa_z) \\ &= q^2 g(q/\kappa_z)\end{aligned} \quad (15.3)$$

where f and g are scaling functions.

Figure 15.1 plots measurements by Hutchings et al. (1972) of $\Gamma_z(\mathbf{q}, t)$ against κ_z for $q = 0$ in FeF_2. The data points can be fitted by

$$\Gamma_z(0, t) = (17 \pm 7)\kappa_z^{2.3 \pm 0.4} \quad \text{meV} \quad (15.4)$$

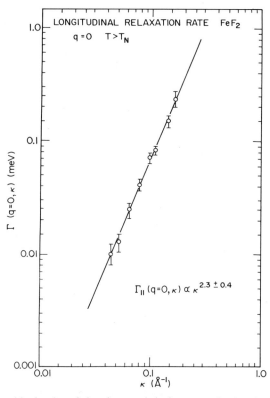

Figure 15.1. Logarithmic plot of the characteristic frequency for longitudinal fluctuations versus the inverse correlation length for FeF_2 at temperatures above T_N and at $q = 0$ (from Hutchings et al. 1972). The solid line represents the best fit to the data of a power law with critical exponent 2.3 ± 0.4.

where the exponent agrees within error with the predicted value of 2.0. Similarly, at the critical temperature Hutchings et al. find that

$$\Gamma_z(q, 0) = (3.4 \pm 0.1)q^{2.1 \pm 0.2} \quad \text{meV} \tag{15.5}$$

where again the measured critical exponent agrees with the predicted value of 2.0. In Figure 15.2 the scaling function $f(q/\kappa_z)$ is plotted against q/κ_z (Hutchings et al. 1972). The points do fall very close to two general curves, one for $T > T_N$ and the other for $T < T_N$, so confirming the dynamic scaling theory.

Finally, we will mention the behavior of the spin-wave energy at the zone center, $\omega_0(0, t)$. In FeF$_2$ Hutchings et al. (1972) find that

$$\omega_0(0, t) \sim |t|^{0.72 \pm 0.18} \tag{15.6}$$

while in MnF$_2$, Schulhof et al. (1971) find

$$\omega_0(0, t) \sim |t|^{0.37 \pm 0.02} \tag{15.7}$$

This difference in behavior is presumably due to the energy gap in

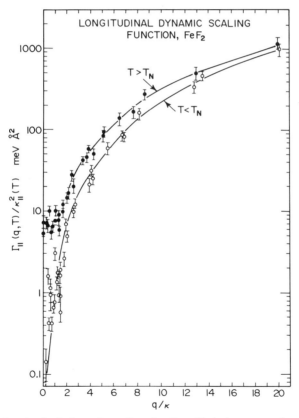

Figure 15.2. Longitudinal dynamic scaling function $f(q/\kappa_z)$ versus q/κ_z for FeF$_2$ at temperatures above and below T_N (from Hutchings et al. 1972).

MnF$_2$, arising from long-range dipolar forces, while in FeF$_2$ it arises from shorter-range effects that fall off more slowly with increasing disorder. However, we know of no quantitative theory of the effect.

15.3. The Ising Dipolar Ferromagnet

In LiTbF$_4$ there exists an Ising-like ferromagnet in which the predominant interactions are dipolar. The long-range nature of these interactions means that we are dealing with a different universality class from the Ising model. Aharony (1973b) has shown that the upper borderline dimensionality is 3. Thus, in three dimensions the Ginzburg–Landau critical exponents should be found, though there are multiplicative correction terms that vary as $\ln(t)$ (Aharony 1973b, Larkin and Khmel'nitzkii 1969). The theory predicts that the magnetization $M(T, H)$ and the correlation length $\xi(T, H)$ vary in the critical region as

$$M(T, 0) = [M(0, 0)B\,|t|]^{1/2}\,|\ln|t||^{1/3} \qquad t < 0 \qquad (15.8)$$

$$\xi(T, 0) = \Gamma_{\pm}^{1/2}\,|t|^{-1/2}\,|\ln|t||^{1/6} \qquad (15.9)$$

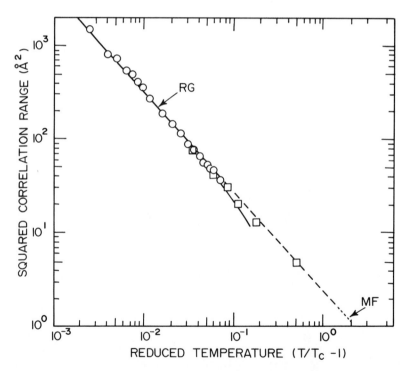

Figure 15.3. Correlation length squared versus the reduced temperature for the dipolar ferromagnet LiTbF$_4$ (from Als-Nielsen 1976b). The full lines shows the prediction of renormalization-group theory and the broken line the prediction of Ginzburg–Landau theory. The fit with theory is excellent and involves no disposable parameters.

Aharony and Halperin (1975) have further shown there are exact relations between amplitudes (B in Eq. 15.8, Γ_+ and Γ_- in Eq. 15.9) near critical points of borderline dimensionality. Using these relations and specific-heat data, Als-Nielsen (1976b) showed that Eq. 15.9 could be fitted to the experimental neutron-scattering data with no disposable parameters. The fit is shown in Figure 15.3 and is excellent. The dotted curve at the right represents the predictions of Ginzburg–Landau theory, that is, the predictions without the logarithmic corrections.

Fits to the susceptibility (Als-Nielsen et al. 1974) and to the magnetization (Als-Nielsen et al. 1975) with neutron-scattering data yield $\gamma = 1.13 \pm 0.06$ and $\beta = 0.45 \pm 0.03$, respectively. These values are close to the predicted Ginzburg–Landau values of 1 and of 0.5, respectively, so that theory and experiment are in reasonable accord for dipolar interactions in a ferromagnet. The data for the correlation length and for the magnetization are incompatible with the Ising model, so that the idea that long-range interactions change universality classifications is confirmed.

All ferromagnetic systems have dipolar interactions, so a crossover to this type of behavior should always happen. For materials like iron, with $T_c = 1040$ K, the crossover will, of course, occur much nearer to the critical point than for LiTbF$_4$ ($T_c = 2.9$ K), and only high-resolution experiments will detect its existence.

15.4. Ordering in Alloys

Although the scattering here is nuclear rather than magnetic, we include ordering in alloys in this book because it has a Hamiltonian that corresponds closely to the Ising model.

Suppose that we have an alloy of composition AB. A typical example might be β-brass where A is a copper atom and B a zinc atom. β-Brass has a body-centered cubic lattice; at high temperatures the sites are uniformly occupied and at low temperatures there is an ordered lattice with copper atoms at the cube corners and zinc atoms at the cube-center sites. A critical phase transition separates the ordered and disordered phases.

If a site i is occupied by an A atom we give it a spin variable $S_i^z = +1$ and if it is occupied by a B atom we let $S_i^z = -1$. $J_{AA}(\mathbf{r}_{ij})$ is the energy of the configuration if sites i and j are both occupied by A atoms, and energies $J_{AB}(\mathbf{r}_{ij})$ and $J_{BB}(\mathbf{r}_{ij})$ are defined analogously. Then the total energy $\mathscr{H}$ of the configuration is given by

$$\mathscr{H} = \tfrac{1}{4} \sum_{ij} J_{AA}(\mathbf{r}_{ij})(1 + S_i^z)(1 + S_j^z)$$
$$+ J_{BB}(\mathbf{r}_{ij})(1 - S_i^z)(1 - S_j^z)$$
$$+ J_{AB}(\mathbf{r}_{ij})[(1 + S_i^z)(1 - S_j^z) + (1 - S_i^z)(1 + S_j^z)]$$

This Hamiltonian leads to the same partition function as for the Ising model (Domb 1974).

In writing down this Hamiltonian we have made two assumptions: first that the energy can be expressed as a sum of pairwise interactions, and second that the occupation of lattice sites does not couple significantly to the lattice, causing strain fields. In fact, there is evidence that three-body interactions are present in actual cases at about a 10% level, relative to two-body interactions (Rácz and Collins 1980), though it is not clear what effect this will have on critical exponents. Since ordering usually takes place only when A and B are atoms with about the same size, strain-field energies would not be expected to be large; however, if such terms exist they will be of long range and will tend to make critical exponents more Ginzburg–Landau-like.

Extensive measurements have been made on two systems: on β-brass (Als-Nielsen and Dietrich 1967, Als-Nielsen 1969, Norvell and Als-Nielsen 1970, Rathmann and Als-Nielsen 1974) by neutron scattering, and on Fe_3Al (Guttman and Schnyders 1969, Guttman et al. 1969) by X-ray scattering. In addition, experiments restricted to measurement of the order parameter have been carried out by neutron scattering on FeCo (Oyedele and Collins 1977) and Fe_3Al (Ahmed et al. 1982) and by X-ray scattering on β-brass (Chipman and Walker 1972).

Table 15.2 summarizes the result of measurements of the order parameter. All the measurements show temperature variations consistent with a power-law variation where the critical exponents β is as given in the table. The consistency of the measured values of β is good, with a weighted mean value of 0.304 ± 0.002. This is clearly lower than the Ising model prediction of 0.326 ± 0.002. Baker and Essam (1971) suggest that the discrepancy is due to the failure to include the lattice compressibility in the Hamiltonian. The lattice parameter is slightly different for the ordered and disordered lattices and the exchange parameters in the Hamiltonian will be sensitive to this change. Baker and Essam estimate that this will effectively renormalize the critical exponent β downwards by

Table 15.2. Measurements of the Critical Exponent β for the Order–Disorder Transition in Alloys

Alloy	β	Reference
CuZn	0.305 ± 0.005	Als-Nielsen et al. (1967)
CuZn	0.315 ± 0.007	Norvell et al. (1970)
CuZn	0.2995 ± 0.0035	Chipman et al. (1972)
CuZn	0.293 ± 0.010[a]	Rathmann et al. (1974)
Fe_3Al	0.307 ± 0.003	Guttmann et al. (1969)
Fe_3Al	0.302 ± 0.009	Ahmed et al. (1982)
FeCo	0.303 ± 0.004	Oyedele et al. (1977)
Weighted mean	0.304 ± 0.002	
Ising model	0.326 ± 0.002	Table 5.1
Compressible Ising model	0.296 ± 0.008	Baker et al. (1971)

[a] Error not given by authors; value quoted is an estimate.

about 0.028 in β-brass, which is somewhat larger than the amount necessary to bring about agreement between theory and experiment.

Figure 15.4 from Rathmann and Als-Nielsen (1974) shows the variation with temperature of the order parameter in β-brass. The neutron and X-ray measurements are in good agreement, but diverge significantly from both the predictions of the Ising model and of the compressible Ising model. Since the measurements agree with the compressible Ising model's value for β, it is surprising that there is no agreement on the magnitude of the order parameter. The reason for the discrepancy is not understood.

Measurements of the critical scattering in both β-brass and Fe$_3$Al show that close to the critical point the scattering cannot be fitted to a

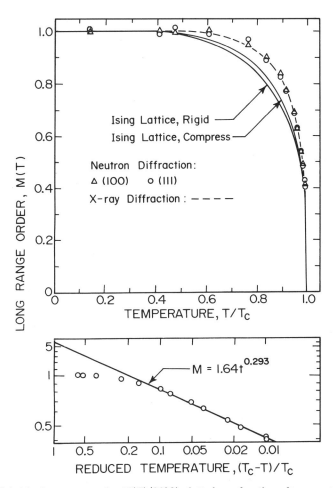

Figure 15.4. The long-range order $M(T)/M(O)$ plotted as a function of temperature (upper panel) and reduced temperature (lower panel) in β-brass (from Rathmann and Als-Nielsen 1974). There is good agreement between neutron and X-ray data, but both diverge significantly from the predictions of the Ising model or of the compressible Ising model.

Table 15.3. Critical Exponents Found From Measurement of the Critical Scattering in β-Brass (Als-Nielsen and Dietrich 1967, Als-Nielsen 1976a) and in Fe$_3$Al (Guttmann and Schnyders 1969)

	β-Brass	Fe$_3$Al	Ising Model
ν	0.65 ± 0.02	0.649 ± 0.005	0.6312 ± 0.0003
γ	1.240 ± 0.015	1.246 ± 0.015^a	1.2378 ± 0.0006
η	0.08 ± 0.07^a	0.080 ± 0.005	0.039 ± 0.002^a

[a] Calculated from the other two parameters using the scaling law $(2-\eta)\nu = \gamma$.

Lorentzian scattering function, just as is found for the two-dimensional Ising model, although the Lorentzian form is not a bad approximation to the scattering. The scattering function $(q^2 + \kappa_1^2)^{-1+\eta/2}$, as given by Eq. 5.38, fits the data better than the Lorentzian function. Als-Nielsen (1976a) finds that the approximate form of Fisher and Burford (1967) probably represents a further improvement beyond this.

Table 15.3 summarizes the critical exponents found in these works from measurements at and above the critical temperature. The critical exponent γ corresponds well to the predictions of the Ising model but both ν and η are slightly larger than expected.

Combination of the scaling laws of Eqs. 5.12, 5.13, and 5.21 yields

$$\gamma + 2\beta = d\nu \qquad (15.10)$$

For a mean experimental value of $\gamma = 1.243 \pm 0.015$ (Table 15.3), and of $\beta = 0.304 \pm 0.002$ (Table 15.2), we predict from Eq. 15.10 that

$$\nu = 0.617 \pm 0.005 \qquad (15.11)$$

This is not consistent with the measured value of 0.649 ± 0.005 (Table 15.3), so that the experimental results violate the scaling laws. Although the critical properties of β-brass and of Fe$_3$Al are close to the predictions of the Ising model, there are significant discrepancies. Normal instinct is to look first for differences in Hamiltonians, rather than to discard much of the theory of critical phenomena, particularly in this case when the magnetic data (Table 15.1), although slightly less accurate, do not point to any problems with the underlying theory. This leads us to look to adding the compressibility and possibly three-body forces to the Hamiltonian. The former process explains the observed values of β and γ (Baker and Essam 1971), while its effect on ν has not been calculated. It is not known what influence the presence of three-body forces might have on the critical exponents, though its effect on the location of the critical point has been estimated (Rácz and Collins 1980).

Suggested Further Reading

Als-Nielsen (1976a)

16

OTHER SIMPLE SYSTEMS IN THREE DIMENSIONS

16.1. Heisenberg Ferromagnet

Just as there was difficulty in finding two-dimensional magnetic systems corresponding to the X–Y model, there is a dearth of work on three-dimensional X–Y materials. Accordingly, we move on to Heisenberg systems, for which materials can be found with the required Hamiltonian. The most extensively studied of these are the ferromagnets EuO and EuS and the antiferromagnet RbMnF$_3$. In this section, work on the two ferromagnets will be described.

EuO and EuS have the cubic rocksalt structure and become ferromagnetic at low temperatures with critical temperatures of 69.2 K and 16.57 K, respectively. Eu^{2+} ions have seven f electrons in the spherically symmetric $^8S_{7/2}$ spin state corresponding to a half-filled shell. This results in a Hamiltonian that follows the Heisenberg model very closely (Passell et al. 1976). Exchange interactions are appreciable to first- and second-nearest neighbors; in EuO both interactions are ferromagnetic with that to second-nearest neighbors about 20% of that to first-nearest neighbors, while in EuS the first-nearest neighbor interaction is ferromagnetic and the second-nearest neighbor interaction is antiferromagnet with magnitude about 50% of the first-nearest neighbor interaction.

The principal cause of departure from the Heisenberg Hamiltonian is dipolar interactions; these must result in a crossover as the critical point is approached. Fisher and Aharony (1973) have estimated the crossover temperature t_c as 0.017 and 0.05 for EuO and EuS, respectively. This is well within the range accessible experimentally and should result in a crossover to an X–Y model. Fisher and Aharony (1973) estimate from a renormalization-group calculation that in the dipolar region γ is increased by 0.03 and η is increased by 0.002 relative to the Heisenberg region. Such a small change in critical exponent is difficult, if not impossible, to detect with current techniques. It is interesting to note that the predicted crossovers are in directions away from the Ginzburg–Landau exponents, which is not what might have been guessed in view of the long-range nature of dipolar interactions.

The crossover is detectable, however, because it results in a divergence of the correlation length for xx or yy correlations at T_c but not of the correlation length for zz correlations. Below T_c the magnetization vector can be aligned by the application of an external magnetic field, so that the parallel (zz) and perpendicular (xx or yy) correlations can be measured separately. Kötzler et al. (1986) have made such measurements of the correlations in single crystals of EuO and EuS with a polarized neutron beam. They show that the two correlation functions are different near T_c as predicted, with only the perpendicular correlation length diverging as T_c is approached.

Unfortunately, these experiments are not accurate enough to enable precise critical exponents to be measured, even though they were made with a sample of the separated isotope ^{153}Eu. Ordinary europium has a large neutron-absorption cross section, which hinders the collection of good experimental data.

The static critical exponents of EuO and of EuS have been measured by Als-Nielsen et al. (1976b) on powder samples with the separated isotope ^{153}Eu. The polycrystalline nature of the sample restricts measurements to the forward direction ($\mathbf{g}=0$). Since we have cubic symmetry and zero applied magnetic field, the critical fluctuations will be isotropic in real space and the measurements will correspond to a sum of the xx, yy, and zz correlations in spin space. The measurements span a reduced temperature range from 1 to 20×10^{-2} and show no crossover effects. The measured critical exponents are quoted in Table 16.1; the agreement with the predictions of the Heisenberg model is excellent.

As well as the static properties, the dynamic critical properties of EuO (Dietrich et al. 1976, Mezei, 1984, Böni and Shirane, 1986) and of EuS (Bohn et al. 1984, Böni et al. 1987a) have also been investigated in detail. Here the effect of dipolar interactions should be easier to observe, since the dynamic critical exponent z is predicted to be 2.48 for the Heisenberg ferromagnet (cf. Eq. 7.13) and 2.0 for dipolar ferromagnets (Riedel and Wegner, 1970, Kötzler, 1983). This large change of exponent should be much easier to observe than the small changes in the static exponents.

The spin waves in EuO have been studied extensively near the critical point by Dietrich et al. (1976). Heisenberg interactions give a spin-wave energy of

$$\omega_c^H(\mathbf{q}, t) = D(t)q^2 \qquad (16.1)$$

at small wavevectors (cf. Eq. 7.9) and dipolar interactions serve to raise this energy to

$$\omega_c(\mathbf{q}, t) = \omega_c^H(\mathbf{q}, t)[1 + 4\pi g\mu_B M \sin^2 \theta_\mathbf{q} (\omega_c^H(\mathbf{q}, t))^{-1}]^{1/2} \qquad (16.2)$$

Table 16.1. Static Critical Exponents of EuO and EuS from the Neutron-scattering Measurements of Als-Nielsen et al. (1976b). The Heisenberg model predictions are from Table 5.1

Exponent	EuO	EuS	Heisenberg Model
β	0.36 ± 0.01	0.36 ± 0.01	0.367 ± 0.004
γ	1.387 ± 0.036	1.399 ± 0.040	1.388 ± 0.003
ν	0.681 ± 0.017	0.702 ± 0.022	0.707 ± 0.003

where M is the magnetization and $\theta_\mathbf{q}$ is the angle between $\mathbf{q}$ and the direction of magnetization (Holstein and Primakoff 1940). In the powder experiment of Dietrich et al. the magnetization has a random direction with respect to $\mathbf{q}$, so that a spectrum of frequencies $\omega_c(\mathbf{q}, t)$ is observed. However, since the magnetization is known at any temperature, it is possible to extract the spin-wave stiffness constant from the data. Figure 16.1 shows the observed temperature variation of the spin-wave excitations in constant-κ scans at $q = 0.2 \text{ Å}^{-1}$. The relatively sharp spin-wave peak at a reduced temperature of 0.15 broadens and becomes lower in energy as the critical point is approached. The broadening cannot be accounted for by the spectrum of frequencies arising from dipolar interactions, and must be due to the intrinsic width of the spin-wave peaks for given $\mathbf{q}$ and $\mathbf{M}$. Dietrich et al. analyzed their results assuming a

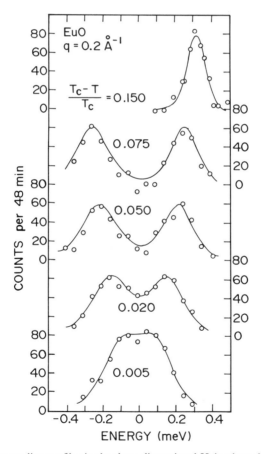

Figure 16.1. Spin-wave line profiles in the three-dimensional Heisenberg ferromagnet EuO for a wave vector of 0.2 Å^{-1} at temperatures just below T_c (from Dietrich et al. 1976). As T_c is approached from below, the spin-wave energy decreases and the line width increases. There is no evidence for an existence of a third line in the profile at zero energy.

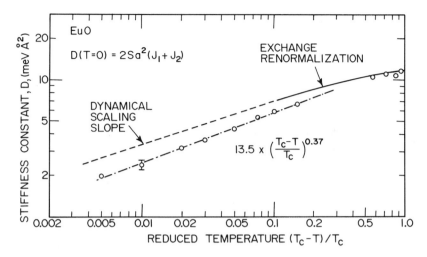

Figure 16.2. Logarithmic plot of the spin-wave stiffness constant against reduced temperature in the Heisenberg ferromagnet EuO (from Dietrich et al. 1976). The data correspond to a critical exponent of 0.37 ± 0.01 while dynamical scaling predicts an exponent of 0.340 ± 0.005. The experimental exponent is sensitive to the model adopted for the damping, and this can lead to systematic errors of the order of 0.03 in the exponent.

Lorentzian shape for the lines at $\pm \omega_0(\mathbf{q}, t)$ with width Γ. That is, they fitted to a spectral weight function

$$2\pi F(\mathbf{q}, t, \omega) = \frac{\Gamma}{(\omega - \omega_0)^2 + \Gamma^2} + \frac{\Gamma}{(\omega + \omega_0)^2 + \Gamma^2} \qquad (16.3)$$

and then identified ω_0 with the characteristic frequency ω_c.

Figure 16.2 shows the results of such an analysis in the form of a log–log plot of the spin-wave stiffness constant $D(t)$ against the reduced temperature t. The plot is linear, showing a critical exponent of 0.37 ± 0.01 for $D(t)$; this is to be compared with the predicted exponent (Eq. 7.12) $\nu - \beta = 0.707 - 0.367 = 0.340 \pm 0.005$ from the static measurements quoted in Table 16.1. The predicted exponent is shown as the dashed line in Figure 16.2; the fact that it always lies above the measured values should be disregarded, since it was arbitrarily joined to the low-temperature data. The difference between the predicted critical exponent of $D(t)$ and the measured value looks to be significant, but Dietrich et al. show that in fact it is probably not. The difficulty lies in the assumption of a Lorentzian shape for the scattering (or for $F(\mathbf{q}, t, \omega)$). Halperin and Hohenberg (1969a) have proposed a different spectral weight function,

$$F(\mathbf{q}, t, \omega) = \frac{\pi \Gamma \omega_0^2}{(\omega^2 - \omega_0^2)^2 + \Gamma^2 \omega_0^2} \qquad (16.4)$$

If this form is fitted to the data, the critical exponent for the spin-wave stiffness constant changes from 0.37 to 0.33. This change arises because for small damping the two forms give virtually the same value of ω_0, while, if the damping is large, the value of ω_0 is different; this changes the slope of Figure 16.2. There is also a problem in that dynamic scaling gives predictions for ω_c, as defined by Eq. 7.3, while the experimentalists determined the spin-wave energy ω_0. These two quantities are not the same, except in the limit as the damping tends to zero. The difference is particularly marked for the double-Lorentzian form, where $\omega_c > \omega_0$ so that the critical exponent is overestimated. In Section 18.2 we show evidence that for cobalt the Halperin–Hohenberg spectra weight function fits the experimental data better than the double-Lorentzian.

These difficulties lead to systematic errors in the determination of the critical exponent of ω that are of the order of 0.03 (Passell et al. 1972). It must be concluded that there is no disagreement between experiment and dynamic scaling.

There is no crossover effect between Heisenberg and dipolar regions in the spin-wave data because the Heisenberg and dipolar terms in Eq. 16.2 ($D(t)$ and M respectively) vary in virtually the same way with temperature.

An interesting feature of Figure 16.1 is that the data can be described in terms of just two spin-wave peaks with frequency $\pm \omega_c$ at all temperatures. Now the spin-wave scattering comes from a response that is purely transverse to the magnetization vector ($\hat{C}^{xx}(\mathbf{q}, t, \omega)$), while at the critical point the response must be isotropic ($\hat{C}^{xx}(\mathbf{q}, t, \omega) = \hat{C}^{zz}(\mathbf{q}, t, \omega)$). There is no sudden change in the integrated intensity ($\hat{C}(\mathbf{q}, t)$) close to T_c and there is nothing in the scattering that can obviously be linked to a longitudinal response below T_c. The data are most easily described if it is assumed that $F^{xx}(\mathbf{q}, t, \omega) = F^{zz}(\mathbf{q}, t, \omega)$, though there is no theoretical explanation of such a situation.

The dynamic critical exponent z can most easily be determined by measuring the wavevector dependence of the scattering at the critical temperature. Dynamic scaling theory predicts (through Eq. 7.6) that the response will have the same shape for all wavevectors q and that the response will have a characteristic width that varies as q^z. Dietrich et al. (1976) measured this width by fitting the line shape to a Lorentzian function convoluted with the resolution function. More recently, Böni and Shirane (1986) and Böni et al. (1987b) have remeasured the line shapes and analyzed them without assuming a Lorentzian shape. Results from Böni and Shirane (1986) are shown in Figure 16.3 in the form of a log–log plot of the half-width at half maximum of $F(\mathbf{q}, 0, \omega)$ against the wavevector $\mathbf{q}$. The systematic differences between the results of Böni and Shirane and of Dietrich et al. arise from the assumption by the latter authors of a Lorentzian line shape. In addition, Mezei (1984, 1986) has used the neutron spin-echo method to extend the data to lower energies by almost two orders of magnitude. He finds, somewhat surprisingly, that his data are consistent with a correlation function in real time, $C(\mathbf{q}, t, \tau)$, that decays exponentially in time. This Fourier transforms to a Lorentzian function, which is contrary to the findings of Böni and Shirane and of Böni et al. at larger wavevectors (and hence violates dynamic scaling). It is also contrary to the theoretical expectations of a function that is squarer-shaped than a Lorentzian (Lovesey and Williams 1986, Folk and Iro 1985). However, if Mezei's data are included on the same logarithmic plot for the characteristic width as the data of Böni and Shirane, one finds a linear variation over four decades of energy with a slope of 2.5. This is shown in Figure 16.3 and the results are as predicted for a Heisenberg ferromagnet (Eqs. 7.13 or 7.14).

This is a further puzzle since, as described earlier, one would have expected a crossover to a dipolar region with $z = 2$ at short wavelength. Kötzler (1983) predicts that the crossover should occur at about

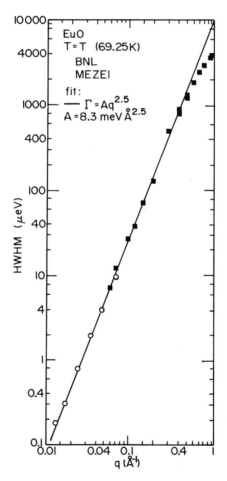

Figure 16.3. Logarithmic plot of the line width of the critical scattering at T_c versus wavevector **q** in EuO (from Böni and Shirane 1986). The data follow a variation with $q^{2.5}$ over four decades of energy. There is no evidence that for wavevectors less than 0.15 Å^{-1} there is a crossover to a dipolar regime with critical exponent 2.0.

$q = 0.15 \text{ Å}^{-1}$. Clearly this does not actually happen. Kötzler suggests that this crossover only occurs for longitudinal fluctuations while Mezei's measurements are primarily of transverse fluctuations because the specimen was in a magnetic field along q. He postulates a dynamic decoupling of the transverse fluctuations at long wavelength.

Bohn et al. (1984) and Böni et al. (1987a) have measured the line width in EuS at the critical temperature. The line width at a given wavevector is 2.9 times narrower than in EuO and this makes measurement more difficult. Figure 16.4 from Böni et al. (1987a) plots logarithmically the line width against the wavevector. The crossover from dipolar

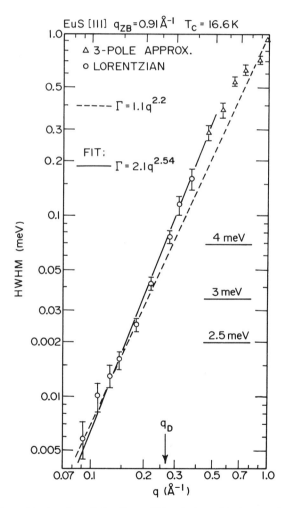

Figure 16.4. Logarithmic plot of the line width of the critical scattering at T_c versus wavevector **q** in EuS (from Böni et al. 1987a). The data show some evidence for the predicted crossover from Heisenberg behavior with $q > 0.27 \text{ Å}^{-1}$ and critical exponent 2.5 to dipolar behavior for $q < 0.27 \text{ Å}^{-1}$ with critical exponent 2.0.

to Heisenberg behavior should be at a wavevector of $0.27\,\text{Å}^{-1}$ and inspection of Figure 16.4 shows that the data are not inconsistent with a cross over from a value of z around 2.5 in the Heisenberg region ($q > 0.27\,\text{Å}^{-1}$) to a value of around 2.2 in the dipolar region ($q < 0.27\,\text{Å}^{-1}$).

16.2. Heisenberg Antiferromagnet

It is fortunate that in $RbMnF_3$ there exists a simple-cubic antiferromagnet with Hamiltonian corresponding very closely to the nearest-neighbor Heisenberg model. Even the ubiquitous dipolar interactions become irrelevant variables in a renormalization-group treatment (Aharony 1973a), so that the Hamiltonian for $RbMnF_3$ is probably closer to that of an ideal model than is that of any other magnetic system. Moreover, with a critical temperature as high as 83.0 K the characteristic frequencies are large enough to be measured in the critical region by neutron-scattering techniques without resolution effects being excessively large.

Extensive measurements of both the static and the dynamic critical properties of $RbMnF_3$ have been made by Tucciarone et al. (1971). We begin by describing their measurements of the static properties. This initially involves using the static approximation (Eqs. 11.22 and 11.23); Tucciarone et al. were able to improve on this by using their later dynamic data to estimate the corrections to the static approximation through explicit self-consistent evaluation of the integral in Eq. 11.18.

Figure 16.5 shows the critical scattering as measured by a two-axis spectrometer at a temperature very close to the critical temperature. The scattering angle is set at the antiferromagnet Bragg peak position, and the crystal is rotated through the Bragg peak position. The measurements were fitted to Eq. 5.38 with ψ set equal to unity, that is, to a static correlation function that varies as $(\kappa_1^2 + q^2)^{-1+\eta/2}$. If the parameter η is set equal to zero, the measurements cannot be fitted to the assumed form of the correlation function (after convolution with the experimental resolution function). The two solid lines in Figure 16.5 correspond to a fit to the wings of the experimental data (which corresponds to $\kappa_1 = 0.00385\,\text{Å}^{-1}$) and to a fit to the center of the data (which corresponds to $\kappa_1 = 0.00185\,\text{Å}^{-1}$). Neither solid line gives a satisfactory description of all the data shown. The dashed curve, which corresponds to $\kappa_1 = 0.00156\,\text{Å}^{-1}$ and $\eta = 0.082$, fits the data throughout. All three fitted curves were made assuming the static approximation. Tucciarone et al. made a number of similar fits at temperatures slightly above the critical temperature and from these they quote a best value for η of 0.067 ± 0.010.

It is to be expected that the data give the most sensitive fit for η right at the critical temperature, where $\kappa_1 = 0$ so that one fewer parameter has to be included in the fit. Tucciarone et al. analyze this data with corrections to the static approximation being applied by making use of

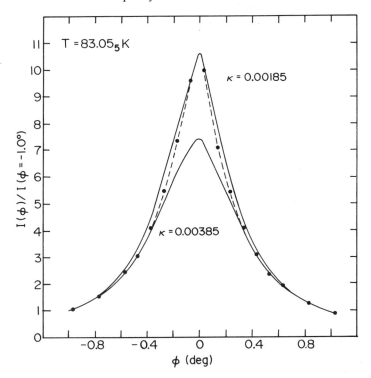

Figure 16.5. The critical scattering observed in the Heisenberg antiferromagnet RbMnF$_3$ at a temperature close to T_N (from Tucciarone et al. 1971). The solid lines are attempts to fit the wings and the central part of the data to a Lorentzian function convoluted with the resolution function. The dashed curve is a best fit if the parameter η is allowed to be nonzero; the fit corresponds to $\eta = 0.082$.

the dynamic data and by assuming dynamic scaling. At "large" wavevectors this fit yields an average value $\eta = 0.053 \pm 0.010$ and at "intermediate" wavevectors the average value is $\eta = 0.044 \pm 0.010$. In the latter of these two ranges the correction to the static approximation reduces η from 0.060 to 0.044.

Tucciarone et al. quote a best value of η as the average of these three separate determinations. This gives

$$\eta = 0.055 \pm 0.010 \tag{16.5}$$

Table 5.1 shows that for the Heisenberg model the value of η is predicted to be 0.037. The uncertainty in this figure is of the order of 0.007 so that, although the experiments give a result that is higher than predicted, the discrepancy is probably not significant. It is noteworthy that this is one of the very few cases (and the only case involving neutron scattering) in which experiment furnishes a critical exponent with precision close to that which can be predicted theoretically for a model system. This is largely because in all other magnetic systems there are

Table 16.2. Comparison of Critical Exponents of RbMnF$_3$ as Measured by Tucciarone et al. (1971) with the Predictions of the Heisenberg Model (from Table 5.1 and Eq. 7.21)

Exponent	Experiment RbMnF$_3$	Heisenberg Model
β	0.32 ± 0.02	0.367 ± 0.004
γ	1.366 ± 0.024	1.388 ± 0.003
ν	0.701 ± 0.011	0.707 ∓ 0.003
ν'	0.54 ± 0.03	0.707 ± 0.003
η	0.055 ± 0.010	0.037 ± 0.009
z	1.46 ± 0.13	1.5

small terms in the Hamiltonian that give rise to doubts about the applicability of model Hamiltonians.

Tucciarone et al. go on to determine the static critical exponents β, γ and ν in RbMnF$_3$ and the results are shown in Table 16.2. In all three cases the relevant parameters vary with reduced temperature in the expected power-law manner. The experimental values for β and γ are lower than predicted by theory. If these discrepancies were real, they would highlight a serious problem.

Tucciarone et al. (1971) have also made extensive measurements of the dynamic properties of RbMnF$_3$. Below the critical temperature, the dynamics show spin-wave excitations. At small wavevectors the energy of the spin wave is $c(t)q$ (cf. Eq. 7.17) and the damping is smaller, and so is a less important correction factor, than in the ferromagnetic case. Figure 16.6 shows the variation of $c(t)$ with temperature on a logarithmic scale. The measurements follow a power law with critical exponent 0.27 ± 0.015. Hydrodynamic theory gives this exponent as $\nu'/2$ (cf. Eq. 7.20, where we made the additional assumption that $\nu' = \nu$). This leads to the

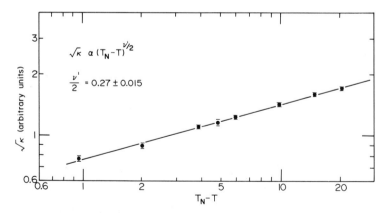

Figure 16.6. Logarithmic plot of spin-wave constant $c(t)$ versus temperature for the Heisenberg antiferromagnet RbMnF$_3$ below T_N (from Tucciarone et al. 1971). Dynamic scaling predicts that $c(t)$ varies as $\sqrt{\kappa_1}$, which leads to a predicted critical exponent of $\nu'/2$.

results $v' = 0.54 \pm 0.03$, which implies either that hydrodynamic theory has failed, or that $v \neq v'$, which is a violation of static scaling.

Figure 16.7 shows the line shapes of the dynamic response at the critical temperature as a function of wavevector q. The scattering cross section should vary as (Eqs. 5.34, 7.5, 7.6, 7.8)

$$\hat{C}(\mathbf{q}, 0, \omega) \sim q^{\eta-2}\omega_c^{-1}F(\omega/\omega_c) \tag{16.6}$$

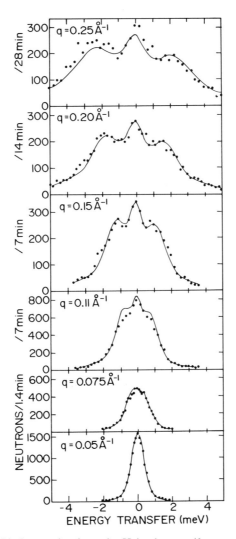

Figure 16.7. The critical scattering from the Heisenberg antiferromagnet RbMnF$_3$ at the critical point for six different wavevectors q (from Tucciarone et al. 1971). Dynamic scaling predicts the same line shape at all wavevectors with the width varying as $q^{1.5}$. The solid lines are fits to this prediction convoluted with the resolution function. There is a peak in the scattering at energy transfer zero as well as peaks at $\pm\omega$ corresponding to damped spin waves.

with
$$\omega_c \sim q^z \tag{16.7}$$

The solid lines in the figure are a fit to this form with $\eta = 0.005$ as previously determined, and with $F(\omega/\omega_c)$ set as the three-peaked function apparent at large wavevectors. The scattering is convoluted with the resolution function and it is the width of the resolution function that turns the observed line shape into a single peak at small wavevectors where ω_c is small. The fit is reasonable with

$$z = 1.4 \pm 0.1 \tag{16.8}$$

This is in satisfactory agreement with the prediction that $z = 1.5$ (Eq. 7.21). The three-peaked structure indicates that presence of damped spin waves at the critical temperature, together with a central mode. This is in contrast with the ferromagnets EuO and EuS for which, as we discussed in the previous section, the scattering function shows a single peak at T_c.

Above the critical temperature, the experiments show a dynamic response that is approximately of Lorentzian form at $q = 0$ with width $\Gamma_0(t)$. This width varies with temperature according to a critical exponent of $(1.46 \pm 0.13)\nu$, whereas theory predicts a width of $z\nu$ (Eqs. 7.21, 7.23) so that
$$z = 1.46 \pm 0.13 \tag{16.9}$$

This is in satisfactory agreement with the prediction $z = 1.5$ and with the experimental result given in Eq. 16.8.

To summarize, the data for RbMnF$_3$ provide theory with a searching test. There is agreement for many critical properties, but there is a sharp disagreement over the temperature variation of the spin-wave energy. There are small discrepancies, which may not in fact be significant, over the critical exponents β, γ, and η. For ν and z, experiment and theory are in close agreement.

16.3. Transitions with SO(3) Universality Class

If one considers a simple hexagonal lattice with Heisenberg Hamiltonian and antiferromagnetic nearest-neighbor interactions in the hexagonal plane and either ferromagnetic or antiferromagnetic interactions along the c axis, then one can show that for a classical system the lowest energy state corresponds to a triangular arrangement of spins at 120 degrees to each other. This triangle can have its plane oriented in any direction in space, since the spins within the triangle can be rotated together about any axis without change of energy. Figure 16.8 shows the structure for a triangle in the *ab* plane. The nearest neighboring spin to any given spin along the c axis is either ferromagnetically or antiferromagnetically aligned to that spin.

It is clear that the order parameter for this arrangement cannot be represented as a single three-dimensional vector even though the Hamiltonian corresponds to the Heisenberg model. Kawamura (1985,

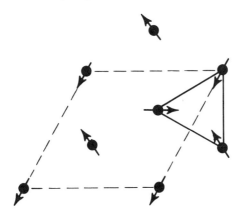

Figure 16.8. The triangular antiferromagnetic structure. The broken lines indicate the unit cell and the continuous lines show one of the equilateral triangles of neighboring atoms that make up the structure. Spins are oriented at 120 degrees to each other. For artistic convenience, the plane of the spin vectors is shown as the same plane as the plane of the lattice.

1986) shows that the order parameter has the symmetry of the three-dimensional rotation group SO(3). This introduces a new universality class. Kawamura estimates the critical exponents to be $\gamma = 1.1$, $\nu = 0.53$, $\beta = 0.25$, and $\alpha = 0.4$ (assuming scaling to hold).

There are a number of materials with this crystal structure and Hamiltonian, though invariably there is a small anisotropic term also present in addition. For most of these materials, an anisotropy confines the spins to within the plan normal to the c axis; this reduces the symmetry from SO(3) to $Z_2 \times S_1$. Some materials have exchange parameters along c that are much greater than in the plane, so as to form quasi-one-dimensional systems (e.g., TMMC or $CsMnBr_3$). Others have the exchange parameter along c must less than in the plane, so as to form quasi-two-dimensional systems (e.g., VCl_2 or VBr_2).

The only material with SO(3) symmetry that has been studied extensively by neutron scattering techniques in the critical region is VCl_2 (Kadowaki et al. 1987). The exchange interactions along the c axis are only 0.6% of the exchange interactions in the plane, so that the behavior would be expected to be as for a two-dimensional SO(3) system far from the critical point with a crossover to three-dimensional SO(3) behavior as the critical point is approached. Very close to the critical point ($T_N = 35.9$ K) there will be a second crossover caused by the small Ising-like anisotropy. These crossovers are observed at approximately $t = 0.1$ and $t = 0.01$, respectively.

In the region with $t > 0.1$, the scattering is independent of q_z, which confirms the two-dimensional nature of the fluctuations. As is usually the case, the experiments are not very sensitive to the value of the critical exponent η, but indicate that $\eta = 0.4 \pm 0.1$.

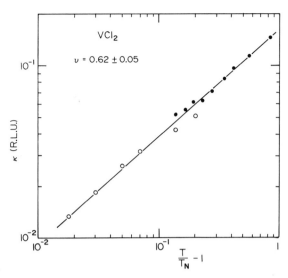

Figure 16.9. Logarithmic plot of the inverse correlation length against reduced temperature in the paramagnetic region for VCl_2, a material whose order parameter has SO(3) symmetry (from Kadowaki et al. 1987).

Near $t = 0.1$, there is a crossover in the nature of the critical scattering from two-dimensional to three-dimensional behavior. This crossover is not apparent, however, in the magnitude of the correlation length or of the susceptibility. Figure 16.9 shows a logarithmic plot of the variation of the reciprocal correlation length κ_1 with reduced temperature. The plot is linear right through the crossover at $t = 0.1$ with a slope corresponding to $v = 0.62 \pm 0.05$. Similar plots show that in both regions $\gamma = 1.05 \pm 0.03$ and $\beta = 0.20 \pm 0.02$. The agreement with the predictions of Kawamura for the SO(3) model of 0.53, 1.1, and 0.25, respectively is not too bad. It is perhaps noteworthy that none of our standard models gives a set of critical exponents close to these values; this strengthens the proposition that the critical properties belong to a new universality class, SO(3).

An analogous situation occurs in the critical transition in $CsMnBr_3$, although here the anistropy is sufficiently large that the spins are effectively confined to the $X-Y$ plane in the critical region. This lowers the symmetry from SO(3) to $Z_2 \times S_1$. Mason et al. (1987) have shown that this critical phase transition has an exponent β of 0.22 ± 0.02. As with VCl_2, this exponent does not correspond to any standard model and is close to the value of 0.25 ± 0.02 predicted by Kawamura.

Suggested Further Reading

Als-Nielsen et al. (1976)
Als-Nielsen (1976a)
de Jongh and Midiema (1974)
Tucciarone et al. (1971)

17

MULTICRITICAL POINTS

17.1. Tricritical Points

There are relatively few detailed investigations of multicritical phenomena by neutron-scattering techniques. An example of measurements around a tricritical point is the work of Bongaarts and de Jonge (1977) on $CsCoCl_3 \cdot 2D_2O$. The phase diagram is shown in Figure 17.1. It is of the same form as Figure 8.3. A practical complication is that near the first-order phase transition line the field induces a ferromagnetic moment that is sufficiently large that demagnetizing fields cannot be neglected in comparison with the applied field H. This actually gives rise to a mixed-phase region.

Bongaarts and de Jonge measure first the variation of the order parameter (the staggered magnetization) close to the critical point at $H = 0$, and find a power law with critical exponent $\beta = 0.298 \pm 0.006$. Measurements of the exchange parameters indicate that the coupling in $CsCoCl_3 \cdot 2D_2O$ is about two orders of magnitude stronger in one direction than in either of the other two directions, but the critical exponent shows no signs of crossover behavior for reduced temperatures between 10^{-1} and 10^{-3}. The experimental value of β is close to, but significantly lower than, the value 0.326 expected for the three-dimensional Ising model (cf. Table 5.1).

Near the tricritical point in the critical phase transition region, Bongaarts and de Jonge measure both the antiferromagnetic order parameter (the staggered magnetization) and the magnetization as a function of field at fixed temperature. Figure 17.2, taken from their paper, shows how these plots give critical exponents near the tricritical point (tricritical exponents) that cross over to ordinary critical exponents as we move away from the tricritical point. This is as was predicted in Chapter 8 (cf. Eq. 8.12).

The tricritical exponent of the order parameter is found to be 0.15 ± 0.02. This is lower than the Ginzburg–Landau value of 0.25 (Eq. 8.13), but such a difference is perhaps not surprising when at the ordinary critical point ($H = 0$) Ginzburg–Landau theory predicts an exponent of 0.50 and experiment gives 0.298. The idea that exponents are different at tricritical points than at ordinary critical points is confirmed.

If we expand α_4 as a linear function about the tricritical point (just as α_2 is expanded in Ginzburg–Landau theory in Eq. 2.2) we expect, from Eq. 8.14, that the jump in the staggered magnetization will vary as the

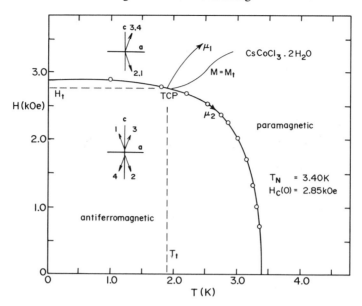

Figure 17.1. Magnetic phase diagram of $CsCoCl_3 \cdot 2D_2O$ for a magnetic field applied along the a direction (from Bongaarts and de Jonge 1977). There is a tricritical point at 1.85 K and 2.70 kOe.

square root of the distance from the tricritical point. Bongaarts and de Jonge measure this critical exponent to be $\beta_1 = 0.3 \pm 0.1$.

There are other critical exponents associated with tricritical points and Bongaarts and de Jonge also determine a number of these; for details the reader is referred to the original paper.

Another example of a magnetic tricritical point is in $FeCl_2$ (Birgeneau et al. 1974, Birgeneau 1975). The measurements are less extensive than those for $CsCoCl_3 \cdot 2D_2O$ but do show tricritical behavior with $\beta_1 = 0.19 \pm 0.02$.

Hirte et al. (1984) have shown that in real systems the situation may become more complicated than just exhibiting a tricritical point, in that an extra, intermediate phase may appear, which complicates the interpretation.

17.2. Bicritical Points

$CsMnBr_3 \cdot 2D_2O$ is a material that shows a bicritical point. In zero applied field it is a weakly anisotropic antiferromagnet with a single easy axis and with a critical temperature of 6.3 K. If a magnetic field is applied along the easy axis, there is a phase transition to a spin–flop phase; the phase diagram is like the phase diagram shown in Figure 8.5 with a bicritical point where the three phases meet at a temperature of 5.26 K and a field of 26.6 kOe.

Basten et al. (1980) have examined these phase transitions in detail by

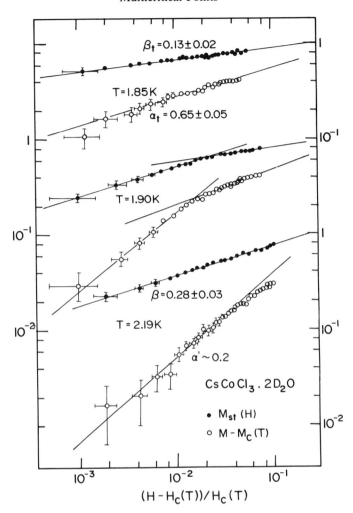

Figure 17.2. Logarithmic plot of the field dependence of the magnetization and the staggered magnetization of $CsCoCl_3 \cdot 2D_2O$ at three different temperatures (from Bongaarts and de Jonge 1977). At $T = 2.19$ K there is ordinary critical behavior, at $T = 1.85$ K tricritical behavior, and at $T = 1.90$ K a crossover between these two forms.

neutron-scattering techniques. Figure 17.3 shows the phase diagram in the vicinity of the bicritical point, with the inset showing the behavior very close to the bicritical point on an expanded scale. The reader is referred to the paper of Basten et al. (1980) for details of the measurement of static critical exponents in the bicritical region.

17.3. Lifshitz Points

A classic case of a material exhibiting a Lifshitz point is MnP (Shapira 1984). The phase diagram is generally similar to that shown in Figure 8.6,

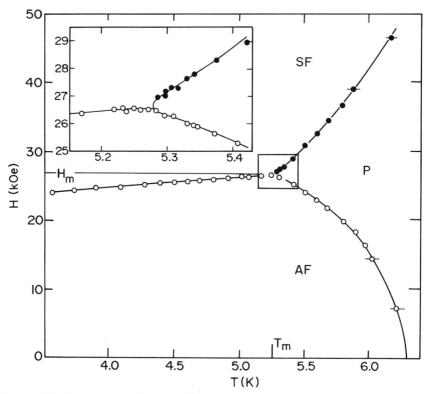

Figure 17.3. Magnetic phase diagram of CsMnBr$_3$·2D$_2$O from Basten et al. (1980). There is a bicritical point at 5.26 K and 26.6 kOe. The inset shows an expanded version of the diagram near the bicritical point.

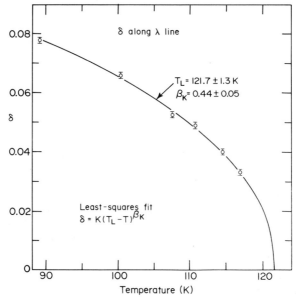

Figure 17.4. Temperature dependence of the incommensurate wavevector δ in MnP as the Lifshitz point is approached (from Moon et al. 1981).

138

though the regions with the commensurate and incommensurate phases are interchanged and the dashed line in the upper part of the diagram has negative rather than positive slope.

The incommensurate phase is helicoidal and is generally referred to as the "fan phase". Figure 17.4 shows data of Moon et al. (1981) of the variation of the incommensurate wavevector ($\delta = aq/2\pi$, with a equal to the nuclear lattice spacing) as the Lifshitz point is approached along the fan–paramagnetic phase boundary. The measurements were fitted to a power-law behavior:

$$q \sim (T_L - T)^{\beta_k}$$

In Figure 17.4 there is a good fit if $\beta_k = 0.44$; later unpublished work by Moon (quoted by Shapira 1984) indicates a best value of the critical exponent β_k to be

$$\beta_k = 0.480 \pm 0.013$$

Ginzburg–Landau theory gives $\beta_k = 0.5$, while application of the renormalization group indicates $\beta_k = 0.54$ (Mukamel 1977).

Suggested Further Reading

Bongaarts and de Jorge (1977)
Basten et al. (1980)
Pynn and Skjeltorp (1984)
Lindgard (1978)

18

CRITICAL PHASE TRANSITIONS IN MAGNETIC METALS

18.1. Introduction

There are many metals that show magnetic critical phase transitions. We divide these into two groups; first, transition metals, in which the magnetism arises from the presence of electrons in d orbitals, and second, rare-earth and actinide metals, in which the magnetism arises from the presence of electrons in f orbitals.

The transition-metal magnets that have been studied involve the elements Cr, Mn, Fe, Co, or Ni. Only in very few cases, such as the Heusler alloys, does it seem that the magnetic electrons can be regarded as localized. In metals like chromium the 3d electrons seem definitely to be itinerant, while in metallic iron or cobalt the situation is intermediate between localization and itinerancy. Once the magnetic electrons become itinerant, it is clear that our simple model Hamiltonians will not be applicable and that the critical properties need not correspond to any of the universality classes that we have hitherto considered.

The general properties and theory of itinerant electron magnetism are described in a review by Moriya (1985). However, the theory has not been able, so far, to make detailed predictions about behavior in the critical region. Indeed, even the description of the magnetic scattering outside of the critical region is not complete and is often still controversial (Windsor 1978, Moriya 1985, Martinez et al. 1985, Lynn and Mook 1986). This topic is outside the scope of this book, but its difficulty indicates that an understanding of the even more difficult problem of a proper description of the critical behavior is not yet close.

In the rare-earths, such as Gd, Tb, Dy, and Ho, the magnetism arises primarily from 4f electrons that are well localized within the atom. Magnetic interactions between atoms take place via the conduction electrons in a mechanism known as the Ruderman–Kittel–Kasuya–Yoshida (RKKY) interaction (Moriya 1985). This interaction gives the Heisenberg model Hamiltonian, though the range of interaction is intermediate rather than short. The Hamiltonian of most rare-earth metals contains, in addition, a large single-ion anisotropic term that arises from crystal-field effects.

Cerium and uranium alloys have magnetism that arises predominantly from f electrons but that does not follow the pattern for the later rare-earth or actinide elements. This is because these are the first elements of the series (rare-earth or actinide) and the f electrons cannot be regarded as well localized within the atom as is the case for later members of the series. The appropriate description of cerium or uranium alloys is the least understood of the metallic magnets.

18.2. Iron, Cobalt, and Nickel

The critical properties of iron, cobalt, and nickel have been studied extensively both by neutron-scattering techniques and by other methods. The critical temperatures are 1044, 1388, and 631 K, respectively, which are all above room temperature and easily accessible experimentally. Iron has a body-centered cubic structure at the critical temperature while both cobalt and nickel are face-centered cubic.

At temperatures well below T_c, the magnetic excitations observed at long wavevectors correspond to well-defined spin-wave modes with an energy that follows Eq. 7.9, that is, an energy that varies as Dq^2 (Collins et al. 1969, Glinka et al. 1977, Minkiewicz et al. 1969, Mitchell and Paul, 1986). Dipolar contributions to the spin-wave energy are negligible in the critical region at wavevectors where triple-axis and double-axis measurements have been made (though, as we shall see later, they should not be negligible at the shorter wavevectors accessible to neutron spin-echo techniques).

Since the order parameter is the magnetization, which is a three-dimensional vector, and the excitations at long wavelength follow the Dq^2 law appropriate for a Heisenberg ferromagnet, it seems more pertinent to compare experimental determinations of the critical exponents with the predictions of the Heisenberg model than of other models. Of course, because of the itinerant nature of the magnetic electrons, it would be no great surprise if the data do not follow this model.

As would be expected from the high values of the critical temperatures, the spin-wave stiffness constant D is quite large for all three metals, so that use of the static approximation in determining critical exponents should lead to appreciable inaccuracy. The only determination by neutron scattering of static critical exponents that goes beyond the static approximation and makes proper correction for inelasticity is the work on cobalt by Glinka et al. (1977). Figure 18.1 shows their log–log plot of χ^{-1} and of κ_1^2 against reduced temperature. The plots are linear over a decade of temperature.

Before accurate measurements were available for the inelasticity of the scattering, Bally et al. (1968a,b) measured γ and ν for Co and Fe by use of the static approximation. As we now know, this procedure is doubtful, though at the time when the measurements were taken this was less clear. However, the values for γ and ν obtained by Bally et al. on cobalt check well with those of Glinka et al. (Fig. 18.1) and the value of γ for iron checks well with magnetic measurements. Thus, it seems likely that the value of ν for iron obtained by Bally et al. is good.

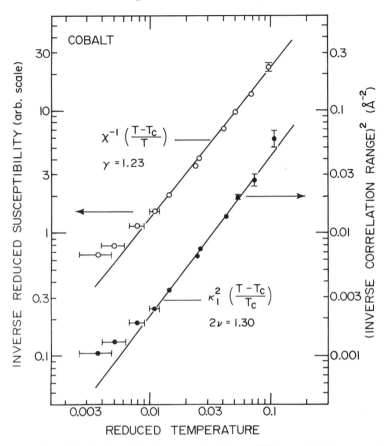

Figure 18.1. Logarithmic plot of the inverse reduced susceptibility and of the square of the inverse correlation length versus reduced temperature in cobalt above T_c (from Glinka et al. 1977).

Table 18.1 gives the determination by neutron-scattering measurements of γ and ν in iron and cobalt and also the determination of γ and β for iron, cobalt, and nickel by magnetization measurements. In every case there is in fact more than one magnetization measurement available; Glinka et al. (1977) list all of these and only a "typical" value is included in Table 18.1.

It is apparent that the measured static exponents do not follow the predictions of the Heisenberg model, or, for that matter, any standard model. The critical exponents of iron and nickel are very similar to each other, while those for cobalt are clearly different. There is no theoretical understanding of these results.

It is interesting to see whether scaling holds for the transition metals. Combining the scaling laws of Eqs. 5.12 and 5.21, we find

$$\gamma + 2\beta - d\nu = 0 \tag{18.1}$$

Table 18.1. The Static Critical Exponents γ, ν, and β for Iron, Cobalt, and Nickel. The Upper Line represents Magnetization Data and the Lower Line Neutron-scattering Data. For Comparison, Predictions of The Heisenberg Model (Table 5.1) are Also Given

	γ	ν	β
Fe	1.33 ± 0.015^a	—	0.389 ± 0.005^b
	1.344 ± 0.018^c	0.679 ± 0.014^c	—
Co	1.20 ± 0.04^d	—	0.42 ± 0.01^e
	1.23 ± 0.05^f	0.65 ± 0.04^f	—
Ni	1.34 ± 0.01^g	—	0.378 ± 0.004^g
	—	—	—
Heisenberg Model	1.388 ± 0.003	0.707 ± 0.003	0.367 ± 0.004

[a] Noakes et al. (1966).
[b] Arajs et al. (1970).
[c] Balley et al. (1968b).
[d] Greissler and Lange (1966).
[e] Rocker and Kohlhass (1967).
[f] Glinka et al. (1977).
[g] Kouvel et al. (1968).

For iron, $\gamma + 2\beta - d\nu$ is 0.02 ± 0.04, and for cobalt it is 0.01 ± 0.12, so that scaling holds in both cases. The test cannot be made for nickel, since ν has not been measured.

Recently, another technique has been introduced for the study of the static exponents β and γ. This involves the study of the depolarization of a neutron beam on transmission through a ferromagnet in the critical region. Stusser et al. (1986) found values of 0.373 ± 0.004 and 0.390 ± 0.004 for β and of 1.34 ± 0.02 and 1.32 ± 0.02 for γ in iron and nickel, respectively. The uncertainty quoted in these values is of the same order as the uncertainty arising from other determinations of β and γ (Table 18.1).

There have been extensive measurements of the critical dynamics of iron, cobalt, and nickel. Below, the critical temperature, spin-wave excitations are observed in all three materials, with energy equal to $D(t)q^2$ (Eq. 7.9). As the temperature is raised, at a small, fixed wavevector q the spin-wave energy reduces and the line broadens. The spin wave becomes overcritically damped before the critical temperature is reached. This is shown for iron in Figure 18.2, which is taken from Collins et al. (1969). The line continues to narrow as the temperature is raised through T_c, reaching a minimum width at a temperature that is above T_c.

The plot is similar to that shown in Figure 16.1 for the Heisenberg ferromagnet EuO at temperatures below T_c. In the discussion of the measurements for EuO it was shown that the critical exponent of the spin-wave stiffness coefficient was slightly sensitive to the nature of the

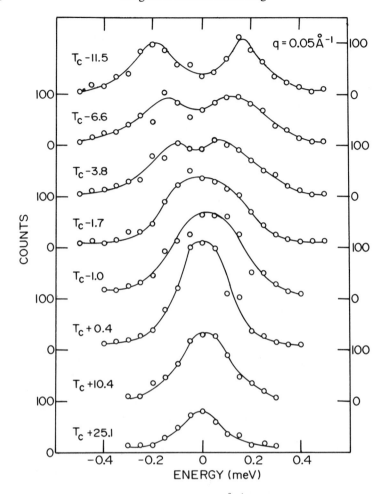

Figure 18.2. Inelastic scattering profiles at $q = 0.05$ Å^{-1} at eight different temperatures in iron (Collins et al. 1969). As T_c is approached from below, the spin wave energy decreases and the line width increases. The spin wave is damped out before T_c is reached. There is no evidence for the existence of a third line in the profile below T_c at zero energy. These results are similar to those found in the Heisenberg ferromagnet EuO (Figure 16.1), but different from the antiferromagnet RbMnF$_3$ (Figure 16.7).

spectral weight function for a damped spin wave. The critical exponent for the Halperin and Hohenberg form (Eq. 16.4) was 0.04 less than the exponent for the double-Lorentzian form. Glinka et al. (1977) investigated the line shape for damped spin waves in cobalt. The upper part of Figure 18.3 shows the result of fitting the two forms to the data; the Halperin and Hohenberg form seems to be clearly superior. The lower part of the figure contrasts the two forms for the spectral weight function for the same values of ω_0 and of Γ. Glinka et al. find that the Halperin and Hohenberg form (after normalization to conform to Eq. 7.2) fits all

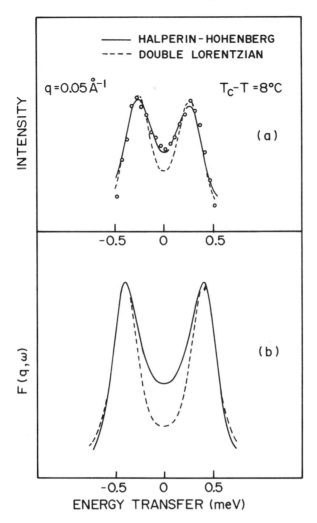

Figure 18.3. The upper panel shows the results of fitting the Halperin–Hohenberg and the double-Lorentzian form to the observed data in cobalt at a temperature 8 K below T_c (from Glinka et al. 1977). The Halperin–Hohenberg form is superior. The lower panel contrasts the two forms when the peak heights and half-widths are matched.

their data below T_c. A similar conclusion was reached by Boronkay and Collins (1973) for spin waves in iron. The adoption of the Halperin and Hohenberg form resulted in a critical exponent for $D(q)$ that was 0.03 less than for the double-Lorentzian form.

As the temperature is raised, the spin waves, at any given wavevector, become overdamped before the critical temperature is reached. This was shown in Figure 18.2 for iron; similar results are found in cobalt, nickel, and EuO (Section 16.1). The three-peak spectrum of the Heisenberg antiferromagnet (Figure 16.7) is not observed. As was noted in Section

16.1, this is puzzling and suggests, unexpectedly, that the longitudinal and transverse correlations may be similar just below T_c.

Figure 18.4 (Minkiewicz et al. 1969) shows the spectral weight function at T_c for iron and for nickel. The shape is the same, though the energy scale differs by a factor of 2 between the two materials. Dynamic scaling predicts that the characteristic width for a given material scales as q^z (Eq. 7.8), though no predictions are made as to the shape of the function.

As was noted in Section 16.1, the derivation of a characteristic width from experimental data over a range of wavevectors necessitates the assumption of a line shape. This line shape is convoluted with the resolution function and fitted to the measurements. For iron and nickel, Collins et al. (1969) and Minkiewicz et al. (1969), respectively, assumed line shapes corresponding to the lines drawn in Figure 18.4. The measurements fitted the predictions of dynamic scaling theory satisfactorily with $z = 2.7 \pm 0.3$ and 2.46 ± 0.25, respectively.

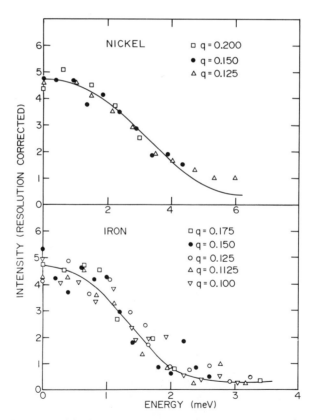

Figure 18.4. Spectral weight function of nickel and iron at T_c for $q = 0.15 \text{ Å}^{-1}$ (from Minkiewicz et al. 1969). Data at wavevectors other than 0.15 Å^{-1} have been scaled to this wavevector using the dynamic scaling laws. The same solid line has been drawn through the nickel and the iron data with the exception that the frequency scale has been changed by a factor of 2.

Glinka et al. (1977) fitted their data to several analytic forms for the spectral weight function and found a Gaussian function to fit best with $z = 2.4 \pm 0.2$. The Lorentzian did not give a satisfactory fit.

Because the characteristic width varies so rapidly with q, all these triple-axis measurements span a rather narrow range of q of less than a factor of 5. Mezei (1984) has used the neutron spin-echo method to extend the data over a much wider range. As for his data in EuO, his measurements suggest a Lorentzian line shape, which is contrary both to the measurements described in the previous paragraph and to theoretical expectation (Lovesey and Williams 1986). However, if Mezei's data are included on the same logarithmic plot as data of Collins et al. (1969) and of Boronkay and Collins (1973), one finds a linear variation over four decades of energy with a slope of 2.48 ± 0.05. This is shown in Figure 18.5, taken from Mezei (1984).

Just as was the case for EuO, Mezei's results are puzzling, firstly because of the Lorentzian line shape and secondly because of the failure of the data to cross over to a dipolar region at small wavevector with $z = 2$ (Kötzler 1983). This crossover is expected to be at a wavevector of about 0.045 Å^{-1} and should be readily apparent in Figure 18.5. ESR and hyperfine interaction measurements indicate that crossover to dipolar behavior occurs in EuO, EuS, Fe, and Ni, but not in Co (Hohenemser et al. 1982).

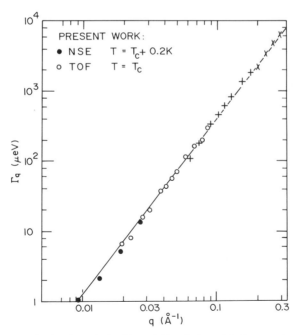

Figure 18.5. Logarithmic plot of the line width of the critical scattering at T_c versus wavevector q in iron (from Mezei 1984). The straight line corresponds to a $q^{5/2}$ power law.

As the temperature is raised above T_c, both the intensity of the critical scattering and the characteristic frequency drop; this same effect was shown for iron in Figure 18.2. The variation of characteristic frequency with wavevector and temperature can be used to check a prediction of dynamic scaling theory. Remembering that in the critical region $\kappa_1 \sim t^\nu$, we can write Eq. 7.7 for an isotropic system as

$$\frac{\omega_c(q,t)}{\omega_c(q,0)} = \frac{f(qt^{-\nu})}{f(\infty)} = F\left(\frac{\kappa_1}{q}\right) \tag{18.2}$$

The ratio of the characteristic frequency at wavevector q and temperature t to that at the same wavevector and at the critical temperature depends only on κ_1/q and not on q and κ_1 individually. Figure 18.6 shows a plot of this ratio of frequencies against κ_1/q for many different values of q and κ_1 in cobalt. It is taken from Glinka et al. (1977) and the measurements do seem to fall on a single curve as predicted by dynamic scaling. The dashed curve is a theoretical prediction of the form of $F(\kappa_1/q)$ by Résibois and Piette (1970) and the solid curve is a fit by Glinka et al. to an empirical function. Similar plots have been made for iron by Als-Nielsen (1970), Boronkay and Collins (1973), and Parette and Kahn (1971).

Above the critical temperature, the measurements should reflect spin diffusion if $q \ll \kappa_1$. Dynamic scaling predicts that the spin-diffusion constant has a critical exponent of $\nu - \beta$ (Eqs. 7.15 and 7.16). Parette

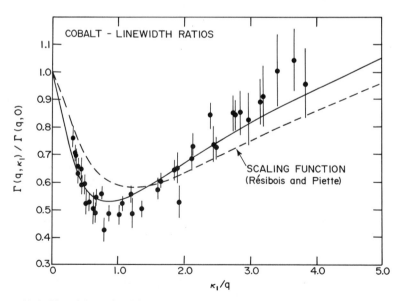

Figure 18.6. Plot of the ratio of the line width at temperature T_c ($t = 0$) against $\kappa(t)/q$ in Co (from Glinka et al. 1977). Dynamic scaling predicts that the ratio should be a function of κ_1/q and this is indeed the case. The dashed line corresponds to a prediction of this functional form by Résibois and Piette (1970).

Table 18.2. Dynamic Critical Exponents for Iron, Cobalt, and Nickel as Determined by Neutron Scattering. For Comparison, Predictions of the Heisenberg Model (Table 5.1) are Also Given

	Line Width at T_c	Spin Wave Stiffness	Diffusion Constant
Dynamic scaling	$2.5 - \eta/2$	$\nu - \beta$	$\nu - \beta$
Heisenberg model	2.48	0.340	0.340
Iron	2.48 ± 0.05^a	0.36 ± 0.03^b	0.38 ± 0.06^c
Cobalt	2.4 ± 0.2^d	0.39 ± 0.05^d	—
Nickel	2.46 ± 0.25^e	0.39 ± 0.04^e	—

[a] Mezei (1984)

[b] Boronkay and Collins (1973).

[c] Parette and Kahn (1971).

[d] Glinka et al. (1977).

[e] Minkiewicz et al. (1969).

and Kahn (1971) have measured this critical exponent in iron; they obtained a value of 0.38 ± 0.06.

Table 18.2 collects together the results of measurements of dynamic critical exponents that have been described. The values are the same for iron, cobalt, and nickel within experimental error in every case. This is unexpected in view of the fact that the static critical exponents of cobalt are significantly different from those of iron and nickel (Table 18.1). The value of the critical exponent for the spin wave stiffness and for the diffusion constant is higher than the value $(\nu - \beta)$ predicted by dynamic scaling, whether $(\nu - \beta)$ be obtained from experimental static exponents (Table 18.1) or from the Heisenberg model. This suggests a failure either of dynamic scaling or of hydrodynamic theory.

18.3. Itinerant Transition Metals: Chromium and MnSi

The critical properties of other transition metals and of alloys involving transition metals have not been studied in as much detail as those of iron, cobalt, and nickel. However, the study of critical properties for itinerant magnets is of special interest because of the expected departure of their behavior from that of the standard models. Such a breakdown of the standard model scheme is found in iron, nickel, and, particularly, cobalt, but it should be even more marked in more itinerant magnets. Two such cases have been studied: chromium and the alloy MnSi.

At low temperatures, chromium exhibits a spin-density wave with a period that is long and is incommensurate with the body-centered cubic lattice. In the simplest antiferromagnetic structure in this lattice, moments point in one direction on atoms at cube corners and in the opposite direction at body-center sites. This gives rise to a magnetic diffraction pattern with Bragg peaks at (hkl) where h, k, and l are integers with

$h + k + l$ odd. Instead of this simple pattern, chromium shows magnetic Bragg peaks close to these expected reciprocal lattice points (**g**) but displaced to a position $\mathbf{g} \pm \mathbf{Q}$, where **Q** is a vector along the x, y, or z direction with a magnitude of about 1/25 of the magnitude of the basic reciprocal lattice vector. The magnetic structure corresponds to the simple antiferromagnetic structure with the magnitude of the magnetic moment modulated sinusoidally with a wavevector **Q**. At temperatures below the spin–flip temperature of 122 K, the moments are aligned in the same direction as **Q** and the spin-density wave is longitudinally polarized. Between 122 K and the Néel temperature ($T_N = 312$ K) the moments are aligned perpendicularly to **Q** and the spin-density wave is transversely polarized (Werner et al. 1967).

The phase transitions at 122 K and at 312 K are both found to be first order in nature. The phase transition at 312 K might be better described as an "interrupted" critical phase transition. This is illustrated by Figure 18.7 (Arrott et al. 1965), which shows the intensity of one of the magnetic Bragg peaks plotted as a function of temperature. The upper plot is for a single-domain crystal, produced by cooling through T_N in a magnetic field, removing the field, and then heating through T_N. The lower plot is for the same peak when the chromium is cooled in the absence of a field so that all domains may be populated. The intensity of the peak is directly proportional to the square of the staggered magnetization (the order parameter). The single-domain crystal shows a linear plot in Figure 18.7, which, if extinction effects can be neglected, would correspond to Ginzburg–Landau theory ($\beta = 0.5$) with a critical temperature of about 41°C. This plot is interrupted by a first-order phase transition at 38.5 K (t ~ 0.008). The long-range nature of the interactions makes it likely that critical properties will follow Ginzberg–Landau theory, though it is not clear why this is interrupted by a first-order phase transition.

The multidomain plot also appears to show an interrupted critical phase transition, though the plot is no longer linear. This interpretation of the data in terms of an interrupted critical phase transition is supported by the presence of critical scattering around the magnetic Bragg peaks (Hamaguchi et al. 1968, Fincher et al. 1981).

The wavevector **Q** of the spin-density wave varies weakly and continuously with temperature from about 21 lattice spacings at low temperatures to about 28 lattice spacings at T_N (Werner et al. 1967). There is no significant change in **Q** on passing through the spin-flop transition nor is there on passing through the Néel temperature (since the critical scattering above T_N peaks about the same wavevector as below T_N (Fincher et al. 1981)).

The critical scattering of neutrons from chromium above T_N is shown in Figure 18.8 (Fincher et al. 1981). These are data taken with a triple-axis spectrometer set for zero energy transfer and scanning along (00ξ) in reciprocal space. At the lowest temperature ($T = 327$ K or $t = 0.05$), the

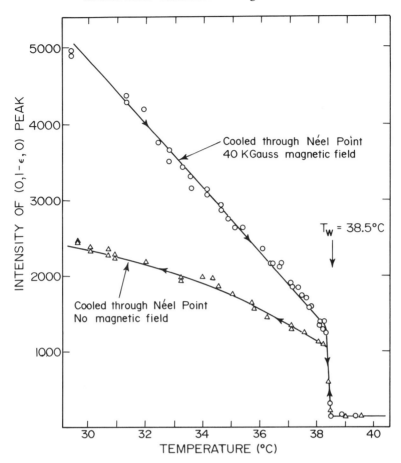

Figure 18.7. The temperature dependence of the intensity of the magnetic Bragg peak $(0, 1 - \epsilon, 0)$ in chromium from 29 to 40°C and back (from Arrott et al. 1965). The plot looks like the intensity profile from a critical phase transition at about 41°C that is "interrupted" by a first-order phase transition at 38.5°C.

critical scattering has peaks at about $(0, 0, 1 \pm 0.03)$ corresponding to the double nature of the Bragg peaks below T_N. The width of these peaks is not resolution-limited and reflects the inverse correlation length κ_1.

As the temperature is raised, the peak changes from being double to being a single peak centered at (001).

Below the Néel temperature, spin-wave branches occur around the magnetic Bragg peaks with energy approximately given by cq, as expected for an isotropic antiferromagnetic structure (Eq. 7.17). However, the branch is extremely steep, with c of the order of 1000 meV/Å (Fincher et al. 1981). This is in fact so steep that it has proved to be impossible to make detailed measurements of the spin-wave properties.

The magnetic properties of chromium are found to be unusually

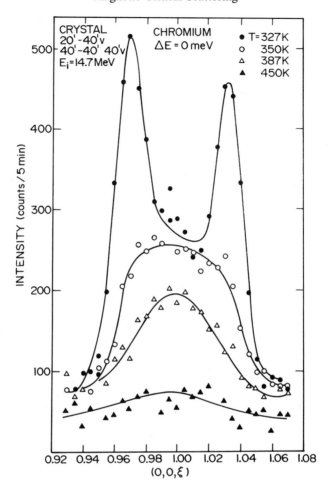

Figure 18.8. The critical scattering from chromium at temperatures above T_N (312 K) (from Fincher et al. 1981). Just above T_N, the critical scattering shows a double peak at $(0, 0, 1 \pm 0.03)$ corresponding to the incommensurate spin-density wave of the ordered structure. At high temperature, the critical scattering shifts to having a single peak centered at (001).

sensitive to the presence of impurities or of strains in the lattice. For instance, the addition of 5% of manganese suppresses the spin-density wave and gives the simple antiferromagnetic structure corresponding to $\mathbf{Q} = 0$ with double the Néel temperature of pure chromium. The critical scattering from this alloy has been studied by Als-Nielsen et al. (1971). They find the same steep spin-wave branches as in pure chromium ($c = 1020 \pm 100$ meV/Å).

The other itinerant magnet whose critical properties have been studied is MnSi. This cubic intermetallic compound has a helical magnetic structure with period 180 Å below its Néel temperature of 29.5 K. The itinerant nature of the magnetism is well established (Ishikawa et al. 1985, Moriya 1985).

In MnSi the magnetic Bragg peaks are found at wavevectors **Q** close to expected ferromagnetic Bragg peaks, unlike chromium in which the peaks are found near expected antiferromagnetic Bragg peaks. The structure is a long-period modulation of a ferromagnetic structure rather than of an antiferromagnetic structure.

As the temperature is raised to T_N from below, the satellite peaks broaden in a direction perpendicular to **Q** and disappear. Above T_N the scattering is peaked at **Q** = 0 and the peak at the low-temperature value of **Q** disappears, so that the scattering looks like critical scattering from a ferromagnet. In chromium, the satellite peaks disappear only at a much higher temperature (Figure 18.8).

In the critical region above T_N, the dynamic properties have been investigated by Ishikawa et al. (1982, 1985). They assume a Lorentzian line shape with width Γ and show that dynamic scaling is obeyed with $z = 2.5$ by plotting $\Gamma q^{-2.5}$ against $\kappa_1 q^{-1}$ (Figure 18.9), just as was done for cobalt in Figure 18.6. The measurements fall on a single curve, as

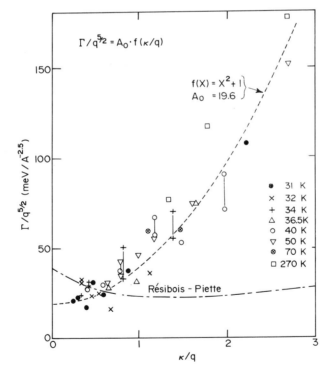

Figure 18.9. Plot of the ratio of the line width at temperature t to the expected line width at T_c ($\sim q^{5/2}$) against $\kappa_1(t)/q$ in MnSi (from Ishikawa et al. 1985). This is quite different from the analogous plot for cobalt shown in Figure 18.6 and from the predictions of Résibois and Piette (1970).

predicted by dynamic scaling, though the curve is quite different from that found for iron or cobalt, or from that predicted by Résibois and Piette (1970).

18.4. Ferromagnetic Transition-metal Compounds: MnP and Pd$_2$MnSn

In this section we cover critical scattering measurements made on two ferromagnetic metallic compounds containing transition metals. First, we look at MnP, a compound that was discussed in the previous chapter in connection with Lifshitz points. In this chapter we look at the ordinary critical phase transition in zero field from ferromagnetism to paramagnetism at a critical temperature of 291 K.

The static critical properties of this phase transition have not been measured by neutron-scattering techniques, but the dynamic properties have been investigated by Minkiewicz et al. (1971) and by Yamada et al. (1987). Minkiewicz et al. (1971) plot the temperature dependence of the spin-wave spectrum at a fixed wavevector in the y direction as shown in Figure 18.10. The spin-wave energy renormalizes to zero as the critical point is approached, just as is the case for the other ferromagnets that we have discussed (EuO, EuS, Fe, Co, and Ni). There is a qualitative difference in the plots, however, since in MnP a central peak is discernible at zero energy transfer while no such peak was apparent in the other cases (Figures 16.1 and 18.2 show analogous plots for EuO and for Fe). This difference is not understood.

Yamada et al. (1987) show that the characteristic frequency near T_c is different in the x and y directions in reciprocal space. There is a corresponding asymmetry in the crystal structure, which is orthorhombic (NiAs type) with c the easy axis and a the hardest axis. This is in contrast to the other ferromagnets that have been discussed so far, where no such asymmetry exists.

Both Yamada et al. (1987) and Minkiewicz et al. (1971) show that the characteristic frequency at T_c along the y direction varies as q^z with z close to the predicted value of 2.5 (Eqs. 7.8 and 7.13). Minkiewicz et al. give $z = 2.57 \pm 0.11$ while Yamada et al. give $z = 2.4 \pm 0.1$ and also find that the fit is slightly improved by assigning a constant term in the equation; that is, writing

$$\omega_{cy} = 0.017 + 150 q_y^{2.5} \quad \text{meV}$$

Along the x axis, Yamada et al. find that

$$\omega_{cx} = 0.017 + 114 q_x^{3.2} \quad \text{meV}$$

Minkiewicz et al. did not make measurements along x.

It is not clear what is the appropriate way to interpret these results, since the anisotropy between x, y, and z can introduce more than one

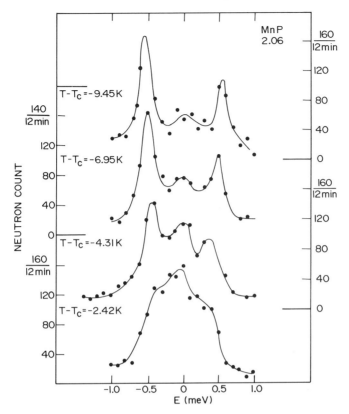

Figure 18.10. Spin-wave profiles in MnP below T_c for $\kappa = (0, 2.06, 0)$ from Minkiewicz et al. (1971). There is a peak in the scattering at zero energy transfer as well as peaks at $\pm\omega$ corresponding to damped spin waves. This is similar to the profile found in RbMnF$_3$ (Figure 16.7), but different from that found in EuO (Figure 16.1) or in iron (Figure 18.2).

crossover region into the dynamics while at the same time MnP is a fairly itinerant ferromagnet (Yamada et al. 1987).

We now turn to the Heusler alloy Pd$_2$MnSn. This is one member of a family of ferromagnetic Heusler alloys in which the magnetism resides predominantly on the manganese atoms. The evidence is that the magnetism is reasonably well localized in the Heusler alloys; in fact, they are some of the most localized metallic transition-metal compounds. They are relatively simple magnetically, since the crystal structure is cubic and there is very little anisotropy.

Pd$_2$MnSn has a critical temperature of 190 K and its spin-wave properties at temperatures below the critical region are as expected for a Heisenberg ferromagnet (Noda and Ishikawa 1976). The critical properties have not been measured extensively, but Shirane et al. (1985) report that $\nu = 0.68$ and $z = 2.5$ and that dynamic scaling holds above T_c with a scaling function close to that predicted by Résibois and Piette

(1970) (shown in Figures 18.6 and 18.9). These results indicate that the Heisenberg model works reasonably well, so confirming the idea that the moments can be considered as localized.

18.5. Rare-earth Metals

The magnetism in the rare-earth metals arises primarily from localized 4f electrons, though smaller magnetic moments (up to above 10%) are associated with the itinerant conduction electrons. The 4f interactions interact via the RKKY mechanism, which is of intermediate range. Unfortunately, only rather limited measurements exist on the critical scattering from rare-earth metals, so that only parts of the picture are available.

Holmium and dysprosium have helically ordered magnetic structures with critical temperatures of 131 and 178 K respectively. As in Cr and MnSi, the magnetic neutron scattering shows Bragg peaks at $\mathbf{g} \pm \mathbf{Q}$, where $\mathbf{g}$ is a reciprocal lattice point and $\mathbf{Q}$ is a wavevector of magnitude small compared with the magnitude of $\mathbf{g}$. As in chromium, $\mathbf{Q}$ varies slowly with temperature and the critical scattering, even just above the critical temperature, is centered at $\mathbf{g} \pm \mathbf{Q}$ (in contrast to MnSi, where it is centered at $\mathbf{g}$). Only the critical exponent β has been measured; for holmium a value of 0.39 (+0.04, −0.03) is found (Eckert and Shirane 1976) and for dysprosium β is 0.39 (+0.04, −0.02) (du Plessis et al. 1983). These values are in exact agreement with each other and with the prediction of Bak and Mukamel (1976) that the system corresponds to a standard model with $d = 3$ and $D = 4$, where β is equal to 0.39.

Terbium has a similar helical magnetic phase, but only over the small temperature region from 219 to 229 K. At lower temperatures it is ferromagnetic and at higher temperatures it is paramagnetic. Dietrich and Als-Nielsen (1967) found $\beta = 0.25$ for the helical-to-paramagnetic phase change, though a more recent measurement by du Plessis et al. (1983) suggested a higher value of β of around 0.33. However, du Plessis et al. found some irreproducibility in their results, and suggest that this might be due to an incomplete first-order phase transition at 219 K. Both works report some residual ferromagnetism in the helical phase.

Dietrich and Als-Nielsen (1968) have measured static critical exponents in terbium and in the paramagnetic phase by neutron scattering. They find critical scattering that peaks at $\mathbf{g} \pm \mathbf{Q}$ and measure critical exponents $\gamma = 1.33 \pm 0.02$ and $\nu = 0.66 \pm 0.02$. These would be more appropriate to the X–Y model ($D = 2$) than to the model with $D = 4$ suggested by Bak and Mukamel (1976).

18.6. Uranium and Cerium Compounds

Uranium and cerium are the first members of the actinide and rare-earth series, respectively. Their magnetic properties are unlike those of the later members of the series because the f electrons are not yet localized within the atom. In fact, the f electrons have effective atomic radii that are as large as if not larger than outer electrons in s, p, or d orbitals. There are more similarities between the magnetic properties of uranium and cerium than between earlier and later members of the actinide or rare-earth series.

Our fundamental understanding of uranium and cerium metals is not as thorough as of the materials that have been discussed previously in this book. Even in insulating compounds the proper description is complicated owing to the presence of mixed-valence effects and of anisotropic exchange. In the metallic state, uranium or cerium compounds may give rise to heavy-fermion properties, though none of the materials that will be described actually fall into this category.

We start this section with the uranium chalcogenides US, USe, and UTe. These are ferromagnets at low temperature with critical temperatures of 177 K, 179 K, and 101 K, respectively. Above T_c the crystal structure is that of sodium chloride; below T_c the moments point along $\langle 111 \rangle$ easy axes and there is a rhombohedral distortion of the lattice. The anisotropy fields at low temperatures are extremely large (over 100 tesla).

Table 18.3 shows the measured critical indices of these material; β and ν were measured by neutron-scattering techniques, and γ and ζ were measured by magnetization techniques. The value of β is quite different for each material and the measured critical exponents do not fit any standard model; indeed, the measured exponents β and ν for USe and for UTe are surprising in that they lie outside the range of values normally encountered in other materials. The scaling law between β, γ and δ (Eq. 5.20) is satisfied but not the scaling law for ν (Eq. 18.1). The low-temperature spin dynamics of these materials are unusual (Buyers and Holden 1985, Hughes et al. 1987 and references therein). US seems to be the farthest from conventional models, with no spin-wave excitations being discernible at low temperatures, while USe shows damped spin waves and UTe has less heavily damped spin waves. The critical dynamics have not been measured for any of these materials and it is clear that there is no fundamental understanding of the measurements.

The uranium pnictides UN, UP, UAs, USb, and UBi have the same crystal strucure as the chalcogenides, but form antiferromagnetic rather than ferromagnetic structures. In the face-centered cubic structure, antiferromagnetism gives rise to a situation known as *frustation*. The energy in each magnetic interaction is minimized when moments coupled by the interaction are aligned at 180 degrees to each other (that is, antiferromagnetically aligned). However, there are triangles of nearest-neighbor bonds because nearest-neighbor atoms may also be nearest neighbors to each other, so that it is not possible to place each moment at 180 degrees to the other two. The system is frustrated.

With just antiferromagnetic nearest-neighbor interactions, the face-centered cubic lattice forms a structure at low temperatures known as the type I structure, but this structure is much more sensitive to small

Table 18.3. Critical Exponents of Some Metallic Uranium and Cerium Compounds with the NaCl Structure. The Measurements for β and ν are by Neutron-scattering Techniques and Those for γ and δ are by Magnetization Techniques

	US[a]	USe[b]	UTe[c,d]	UN[e]	USb[f,g]	CeBi[h]
T_c(K)	177	179	101	50	241	25
Type of ordering	ferro	ferro	ferro	anti	anti	anti
β	0.55 ± 0.05	0.24 ± 0.01	0.291 ± 0.004	0.31 ± 0.03	0.32 ± 0.02	0.317 ± 0.005
γ	1.3 ± 0.1	—	1.33 ± 0.01	1.8 ± 0.1[i]	1.74 ± 0.15	1.16 ± 0.12
δ	3.7 ± 0.5	—	5.23 ± 0.07	—	—	—
ν	—	—	0.84 ± 0.05	0.84 ± 0.05	0.70 ± 0.03[j]	0.63 ± 0.06
$\kappa_{\parallel}/\kappa_{\perp}$	—	—	—	2.9 ± 0.3	5.0 ± 0.5	2.5 ± 0.2

[a] Tillwick and du Plessis (1976).
[b] du Plessis et al. (1982).
[c] Aldred et al. (1980).
[d] Moller et al. (1979).
[e] Holden et al. (1982).
[f] Lander et al. (1978).
[g] Hagen et al. (1988).
[h] Halg et al. (1982a).
[i] γ for UN is estimated from Figure 13 of Holden et al. (1982).
[j] Average of the value for ν_x and ν_y of Hagen et al. (1988).

perturbations than is the structure in nonfrustrated lattices. Several variants on the type I structure are found in the uranium pnictides and ther reader is referred to Rossat-Mignod et al. (1984) for a detailed description.

USb orders in one of these variants that is a noncollinear cubic structure of a class known as a triple-K structure. It has a critical phase transition at 241 K and Figure 18.11 shows the critical scattering at 244.5 K, just above T_c (Lander et al. 1978). The upper section shows the reciprocal lattice in the $1\bar{1}0$ plane; (111) is a nuclear reciprocal lattice point while (001) and (110) are reciprocal lattice points only in the

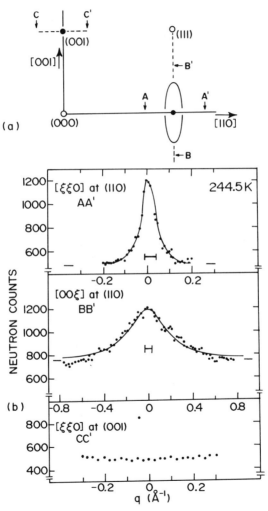

Figure 18.11. Critical scattering from USb just above T_N (from Lander et al. 1978). The critical scattering around (110) is anisotropic in reciprocal space; there is no critical scattering around (001). The data imply that the correlation length is anisotropic in USb.

antiferromagnetic lattice. There is a magnetic Bragg peak at (110) in the ordered lattice, but (001) shows no magnetic Bragg peak because the scattering vector is parallel to the moment direction ($\mathbf{Q}_\perp$ (Eq. 10.19) is zero, so that the cross section is zero). The lower scan of Figure 18.11 shows that there is no measurable critical scattering around (001); this means that there is a large anisotropy in the moment direction even above the critical point. This implies that the system corresponds to an Ising model, because there is no measurable x or y component of the moment though Hagen, Lander and Stirling (1988) indicate that there is an alternative explanation of these facts where the transverse components of the moment give rise to divergences at the critical point which do not contribute to the scattering around (001).

A surprising feature of the critical scattering is that it is anisotropic about (110). It's width is much larger in a scan through (110) perpendicular to the [110] direction (BB' in Figure 18.11) than in the parallel direction (AA' in Figure 18.11). The line width measures κ_1, the reciprocal of the correlation length (Eq. 5.35), so that the extent of the critical fluctuating regions themselves must be anisotropic in real space. The correlation length differs by a factor of 5.0 ± 0.5 in the two directions; this factor is only weakly dependent on temperature, so that the two lengths have the same critical exponent v. As quoted in Table 18.3, Lander et al. (1978) find $\beta = 0.32 \pm 0.02$ for USb.

Mukamel and Krinsky (1976) show that in the absence of anisotropy this material is in the same universality class as the Heisenberg antiferromagnet; Hagen, Stirling and Lander (1988) argue that the presence of anisotropy will not change this conclusion, though there will be cross-over effects at temperatures only infinitesimally below T_c. The measured value of the critical exponent v supports this conclusion, but the measurements of γ do not. Furthermore, measurements of Hagen, Stirling and Lander give a dynamic critical exponent z with value 1.59 ± 0.20, in agreement with the predicted value of 1.5 for the Heisenberg antiferromagnet (Eq. 7.21).

The critical phase transition in uranium nitride at 49.5 K shows broad similarities to that in USb (Holden et al. 1982), though in this case the low-temperature structure is in the single-K class of type I structures. There is an analogous anisotropy to USb, with data similar to that shown in Figure 18.11. The ratio of the parallel and perpendicular correlation lengths is 2.9 ± 0.3, again independent of temperature; this is smaller than the value of 5.0 ± 0.5 found in USb but still very anisotropic. Holden et al. find that the staggered susceptibility and the correlation length in the paramagnetic phase follow the usual form, with a critical power law, but the divergence seems to be at 54.5 K. This temperature is some 10% higher than the Néel temperature $T_N = 49.5$ K, as determined by observation of the variation of the intensity of the magnetic Bragg peaks. No data are given for the intermediate region between 49.5 K and 54.5 K and further work will be needed before the situation becomes clear. As they stand, the measurements point to two critical phase transitions, at 49.5 and 54.5 K, with a phase of unknown nature in the intermediate region. Holden et al. quote $\beta = 0.31 \pm 0.03$ and $\nu = 0.84 \pm 0.05$, while their Figure 13 leads to an estimate of $\gamma = 1.8 \pm 0.1$. The values of ν and of γ are high; Holden et al. suggest that this is because the critical fluctuations are quasi-two-dimensional owing to the anisotropy in correlation lengths. A possible difficulty with this explanation is that USb is even more anisotropic, and yet its value of ν seems to be anomalously low rather than anomalously high.

Now let us turn to UAs, in which as in chromium, there seems to be an "interrupted" critical phase transition (Sinha et al. 1981) at 123.5 K. The ordered phase is a single-K type I structure as in UN (though at 62 K there is a first-order transition to a double-K low-temperature structure). Figure 18.12 shows the critical scattering around (110) (Sinha et al. 1981). There are similarities to the profiles in USb (Figure 18.11) and in UN, but now the critical scattering peaks at $(1, 1, \pm 0.3)$ which suggests that UAs is about to form an (incommensurate) ordered structure with Bragg peaks at $(1, 1, \pm 0.3)$. Before the correlation length actually diverges, however, there is a first-order phase transition to the commensurate type I structure. This is the opposite of the situation in MnSi, where the critical scattering is about a commensurate reciprocal-lattice vector and the ordered state is incommensurate.

Kuznietz et al. (1985) show that the incommensurate phase can be stabilized by replacing 3% of the arsenic atoms by selenium. This causes the incommensurate phase to exist in the temperature range from 113.5 K to 122 K.

Cerium forms an analogous series of pnictides to uranium, with the same crystal structure. The last member of this series, CeBi, has a critical phase transition at 25.35 K with a type I ordered structure. The critical scattering (Halg et al. 1982a,b) shows the same overall features as does the critical scattering of UN and USb. The critical indices are given in Table 18.3; it is remarkable that the critical indices γ and ν differ so much between UN, USb, and CeSb, though β remains the same. The scaling law between the three indices (Eq. 18.1) holds within experimental error for UN and CeSb but not for USb.

CeAs has been less extensively investigated than CeSb or CeBi. It has a critical phase transition at 7.2 K to a type I antiferromagnetic structure; the critical scattering is different from that of the other pnictides in that the anisotropy of the correlation length has the opposite sign (Halg et al. 1982b).

Cerium forms a series of chalcogenides, CeS, CeSe, and CeTe, with the rock-salt structure. These contrast with the ferromagnetism of the uranium chalcogenides by forming type II antiferromagnetic structures with Néel temperatures of 7, 5, and 2 K, respectively (Schrobinger-Papamantellos et al. 1974, Ott et al. 1979). The critical temperatures are low and the exchange energy is comparable with or less than

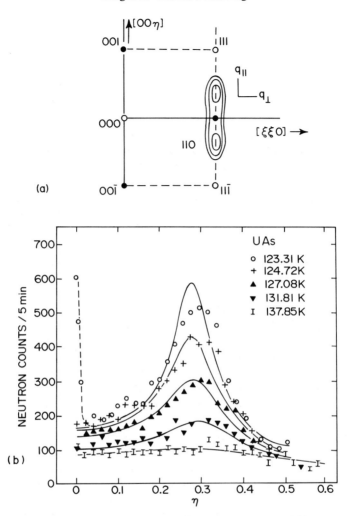

Figure 18.12. Critical scattering in UAs along the line $(1, 1, \eta)$ above and below T_N (from Sinha et al. 1981). The critical fluctuations suggest that an incommensurate magnetic ordered state will form with wavevector close to $(1, 1, 0.3)$; however there is in fact a first-order phase transition at 123.5 K to a commensurate structure with a Bragg peak at (110).

the crystal-field level splittings. The phase transition has been investigated in CeSe and CeTe by Ott et al. (1979) and found to be a normal critical phase transition with $\beta = 0.36 \pm 0.02$ and 0.36 ± 0.04, respectively. This result is in contradiction to the renormalization-group predictions of Mukamel and Krinsky (1976) that the transition should be first order because there is no stable fixed point. Other type II antiferromagnets have also been studied. MnO is found to have an interrupted critical phase transition (Bloch et al. 1975, Boire and Collins

1977) similar to that found in chromium (Figure 18.7) and UAs. The staggered magnetization of MnO follows a typical critical path as the temperature rises, with a critical point of about 123 K, but at 117.5 K this is interrupted by a first-order phase transition. In contrast, NiO also has the type II antiferromagnetic ordering on the rock-salt structure but shows a "normal" critical point (Negovetic and Konstantinovic 1973). The theoretical prediction of a first-order phase transition in type II materials like CeSe, CeTe, and NiO has a solid basis beyond the confines of the renormalization-group picture; it is also predicted by a Ginzburg–Landau approach (Section 139 of Landau and Lifshitz 1969). However, if the degree of interruption of a critical phase transition is very small, its presence may not be apparent experimentally. Historically, the earliest experiments in Cr and MnO were not sensitive enough to detect the presence of the first-order phase transition.

Suggested Further Reading

Collins et al. (1969)
Glinka et al. (1977)
Moriya (1985)
Rossat-Mignod et al. (1984)

19

CRITICAL SCATTERING INVESTIGATION OF DILUTION, PERCOLATION, AND RANDOM-FIELD EFFECTS

19.1. Dilution

In Section 9.1 we saw that theory predicted that dilution need not destroy a critical phase transition. It has no effect on critical indices outside the percolation region if the critical exponent α is less than zero. Of our standard models, only the Ising model in three dimensions has α greater than zero; this case has consequently received most attention.

Both FeF_2 and MnF_2 diluted with zinc have been studied. In these materials it has proved possible to make large samples with the dilutant dispersed homogeneously. Sharp critical phase transitions are observed as expected theoretically. Table 19.1 lists the critical indices measured in the diluted and in the pure systems. The pure systems were described in Chapter 15; FeF_2 is an excellent example of a system following the three-dimensional Ising model and MnF_2 crosses over from three-dimensional Heisenberg to Ising behavior as T_N is approached.

Theory predicts that the Ising critical exponents increase by about 8% on dilution and the experiments generally confirm this, although there is a significant amount of "noise" in the experimental results. The experiments with the lowest quoted uncertainties in the exponents fit the theory best and in no case is there a discrepancy between experimental and theory greater than twice the quoted uncertainties. It is concluded that agreement is satisfactory.

The experimental data for diluted MnF_2 (Mitchell et al. 1986) show no crossover from Ising to Heisenberg behavior as observed in the pure system. This is perhaps not surprising, however, since the diluted Ising exponents are very similar to the Heisenberg exponents (for which pure and diluted are the same), so that a crossover would change the exponents by less than the quoted experimental uncertainty, and hence be beyond detection.

Dilution in the two-dimensional Ising system Rb_2CoF_4 has been investigated by Ikeda et al. (1979) and by Hagen et al. (1987). In this case $\alpha = 0$, which is exactly the boundary value in the application of the Harris criterion. Theory predicts no change in critical exponents and this is exactly what is found to be the case.

Hagen et al. (1987) find that the critical scattering from

Table 19.1. Critical Exponents for the Diluted Three-dimensional Ising Model, Compared with Theory and With the Undiluted Materials[a]

	ν	$\bar{\nu}$	ν'	γ	$\bar{\gamma}$	γ'
$Fe_{0.46}Zn_{0.54}F_2$[b]	—	0.69 ± 0.01	—	—	1.31 ± 0.03	—
$Fe_{0.5}Zn_{0.5}F_2$[c]	0.74 ± 0.03	—	0.72 ± 0.03	1.45 ± 0.06	—	1.43 ± 0.06
FeF_2[d]	—	0.64 ± 0.01	—	—	1.25 ± 0.02	—
$Mn_{0.75}Zn_{0.25}F_2$[e]	—	0.715 ± 0.035	—	—	1.364 ± 0.076	—
$Mn_{0.5}Zn_{0.5}F_2$[e]	0.75 ± 0.05	—	0.76 ± 0.08	1.57 ± 0.16	—	1.56 ± 0.16
MnF_2[f]	0.63 ± 0.02	—	0.56 ± 0.05	1.27 ± 0.02	—	1.32 ± 0.06
Diluted Ising model[g]	—	0.678	—	—	1.34	—
Pure Ising model[h]	—	0.631	—	—	1.24	—

[a] Values plotted under $\bar{\nu}$ or $\bar{\gamma}$ are for cases where it is taken that $\nu = \nu'$ and $\gamma = \gamma'$.
[b] Belanger et al. (1986).
[c] Birgeneau et al. (1983a).
[d] Belanger and Yoshizawa (1987).
[e] Mitchell et al. (1986).
[f] Schulhof et al. (1971).
[g] Jug (1983).
[h] George and Rehr (1984).

$Rb_2Co_{0.7}Mg_{0.3}F_4$ shows the same form for the correlation function and the same critical exponents as for K_2CoF_4 (Cowley et al. 1984), which results were described in Section 14.2. They found $v = 1.08 \pm 0.06$, $\gamma = 1.75 \pm 0.07$ and $\beta = 0.13 \pm 0.02$; these values are in satisfactory agreement with the two-dimensional Ising model predictions of $v = 1$, $\gamma = 1.75$, and $\beta = 0.125$. Although the critical exponents are unaltered by dilution (cf. Table 14.1), the magnitude of the correlation length ξ and of the susceptibility χ at a given reduced temperature are changed quite markedly. The ratios of the correlation length or of the susceptibility above and below T_N for the same $|t|$ are also changed, but Cowley et al. point out that it is not clear whether this is a real effect or is due to a failure of the Tarko–Fisher correlation function.

19.2. Percolation

Measurements have been made of Ising and Heisenberg systems in two and three dimensions. In no case has it proved to be possible to measure the percolation critical indices along the $T = 0$ axis (γ_p, v_p, η_p, etc.),

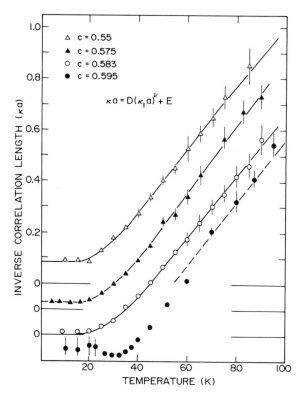

Figure 19.1. The inverse correlation length versus temperature at various concentrations c in $Rb_2Co_cMg_{1-c}F_4$ (from Cowley et al. 1980a). The zeros of the scale are different for each concentration. For $c < p_c = 0.594$ the solid lines are fits to Eq. 9.3. At $c = 0.595$ there is long-range order at temperatures below 30 K.

Table 19.2. Percolation Critical Exponents v_T, γ_T, and ϕ for Ising and Heisenberg systems in Two and Three Dimensions. Theory Predicts $\phi = 1$ for Ising Systems in Any Number of Dimensions, as is Found Experimentally

Material	Model	v_T	γ_T	ϕ
$Rb_2Co_pMg_{1-p}F_4$[a]	Ising $d = 2$	1.32 ± 0.05	2.4 ± 0.1	1.01 ± 0.04
$Mn_pZn_{1-p}F_2$[b]	Ising $d = 3$	0.85 ± 0.10	1.7 ± 0.2	1.0 ± 0.1
$Rb_2Mn_pMg_{1-p}F_4$[c]	Heisenberg $d = 2$	0.90 ± 0.05	1.50 ± 0.15	1.56 ± 0.15
$KMn_pZn_{1-p}F_3$[b]	Heisenberg $d = 3$	0.96 ± 0.05	1.73 ± 0.15	0.91 ± 0.07

[a] Cowley et al. (1980a).
[b] Cowley et al. (1980b).
[c] Birgeneau et al. (1980).

because this requires a more accurate method of determining the concentration of the dilutant than is currently available. Instead, the experiments have been confined to varying the temperature of a given material and determining the thermal properties. The scattering in the paramagnetic region is Lorentzian in form and is fitted to Eq. 5.40. Figure 19.1 shows the inverse correlation length plotted against the temperature for the two-dimensional Ising antiferromagnet $Rb_2Co_pMg_{1-p}F_4$ with $p = 0.55$, 0.575, 0.583, and 0.595. For this material $p_c = 0.594$, so that the largest value of p is above the percolation threshold and shows long-range order at low temperatures. In the paramagnetic region, all the lines have similar slopes at high temperatures and zero slope at low temperatures; there is an excellent fit to the form of Eq. 9.3. The fit in fact enables the concentration p to be determined more accurately than it could be chemically if a theoretical value is assumed for v_p.

Plots similar to those given in Figure 19.1 are found in other materials near the percolation point. The fitted critical exponents v_T and γ_T are given in Table 19.2 for systems corresponding to Ising and Heisenberg models in two and three dimensions. If the theoretical values for v_p and γ_p (Table 9.1) are assumed, then values of ϕ can be deduced from Eq. 9.2 ($v_p = \phi v_T$; $\gamma_p = \phi \gamma_T$). The two values are consistent in every case and the right-hand column of Table 19.2 gives their average value. For the Ising model in any number of dimensions, theory predicts $\phi = 1$, which is confirmed experimentally. For the Heisenberg model in two dimensions, the experiments show clearly that ϕ is greater than unity, though in three dimensions ϕ is not significantly different from 1. Coniglio (1981) predicts for the Heisenberg or the $X-Y$ model that $\phi = 1.43$ in two-dimensional systems and 1.12 in three-dimensional systems.

19.3. Random Fields

We saw in Section 9.3 that theory predicts that the presence of random fields will destroy long-range order in a two-dimensional Ising system.

This has been shown to be the case in $Rb_2Co_{0.7}Mg_{0.3}F_4$ by Yoshizawa et al. (1982) and by Birgeneau et al. (1983b).

Although the random fields destroy the long-range order, there is still short-range order present. The scattering from this short-range order does not follow the usual Lorentzian form (Eq. 5.35) or the slight variants to this form given by Eqs. 5.38 or 5.40. The change in the form of the scattering arises because in these systems there are two sources of short-range correlations, the ordinary critical fluctuations and the fluctuations due to the random field. The resultant correlation function has a term like a Lorentzian and also a term like the square of a Lorentzian function (Pelcovits and Aharony 1985 and references therein). The neutron-scattering experiments verify this form for all diluted antiferromagnetic systems in an external field. This is illustrated in Figure 19.2, where the critical scattering in the static approximation is shown for $Rb_2Co_{0.7}Mg_{0.3}F_4$ (Birgeneau et al. 1983b). The shape is between that of a Lorentzian and a Lorentzian squared; it can be fitted to the sum of a Lorentzian and a Lorentzian squared or to a Lorentzian raised to the power $x/2$ with $x = 3.0$.

The correlation length and the correlation function at $q = 0$ are both found to vary with field H according to simple power laws as given by Eqs. 9.7 and 9.8. At T_N it is found that

$$\nu_H(T_N) = 0.7 \pm 0.2$$

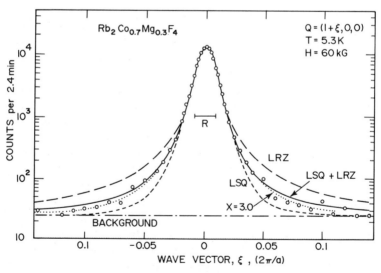

Figure 19.2. Critical scattering in $Rb_2Co_{0.7}Mg_{0.3}F_4$ along $(1 + \xi, 0, 0)$ at 5.3 K with $H = 60$ kilogauss (from Birgeneau et al. 1983b). The applied field and the dilution have created a random-field system. The line shape is between that of a Lorentzian (long-dashed line) and a Lorentzian squared (short-dashed line). It can be fitted to the sum of Lorentzian and Lorentzian-squared terms (solid line) or to a Lorentzian term raised to the power $x/2$ with $x = 3.0$ (dotted line).

Theory predicts $v_H(T_N) = (\gamma/\phi_R)(1 - \eta/2)^{-1}$ (Eq. 9.7), and if α and η are taken from the two-dimensional Ising model and if ϕ_R is assumed to be equal to γ, then the predicted value of $v_H(T_N)$ is 1.14, which is higher than is observed.

v_H and γ_H are found to be functions of T; at the lowest temperature for which results are given (10 K), Birgeneau et al. give a value of 1.6 ± 0.1 for v_H and 3.2 ± 0.2 for γ_H.

In three-dimensional Ising systems, the number of measurements is much larger (see reviews by Wong et al. 1984, Belanger et al. 1984, and Birgeneau et al. 1986). The situation is much less clear than in two dimensions because the measurements in the ordered state turn out to be history dependent. That is to say that the state of the system depends not just on T and H, but also on the path taken to get the system to that particular temperature and field. For example, if one cools to low temperatures in an applied field, a state is found with no long-range order, while if the cooling is in zero field and a field is then applied, long-range order is found. The system is getting hung up in metastable states, something to which Ising systems are especially susceptible because of their lack of intrinsic dynamics. In the two-dimensional system $Rb_2Co_{0.7}Mg_{0.3}F_4$, this metastability only appears at temperatures well below $T_N(0)$, so measurements in the critical region should be reliable.

The onset of history-dependent results makes it hard to answer, even experimentally, questions as to whether the lowest energy state is ordered for the random-field Ising system in three-dimensions.

One measurement that avoids the metastability problem is that of T_N as a function of H. Such measurements have usually been made by techniques other than neutron scattering, so they will not be described here in detail. It will suffice to point out that Eq. 9.6 always seems to provide a good description of the experimental results. Perhaps the most accurate measurements are for $Fe_{0.46}Zn_{0.54}F_2$ (Belanger et al. 1986 and references therein), for which it is found that $\phi_R = 1.42 \pm 0.03$ and $\gamma = 1.31 \pm 0.03$. This is in satisfactory agreement with Aharony's prediction that $\phi_R \simeq 1.1\gamma$ (Eq. 9.5).

Suggested Further Reading

Mitchell et al. (1986)
Birgeneau et al. (1980)
Wong et al. (1984)
Belanger et al. (1984)
Birgeneau et al. (1986)

APPENDIX

Thermal Neutron Scattering Lengths and Cross Section for the Stable Elements and For a Few Selected Isotopes

From the Compilation of V.F. Sears to Appear in the International Tables for Crystallography

The real part of the scattering length $\bar{b}$ is tabulated in femtometers (the imaginary part arises from the absorption). The coherent and incoherent cross sections are defined by Eq. 10.10. The absorption cross section is for incident neutrons with speed 2200 m s^{-1}. Cross sections are given in barns (10^{-28} m^2) for the element as found naturally (N); in a few cases the cross sections are also listed for specific isotopes.

Element	A	Abundance (%)	$\bar{b}$ (fm)	σ_c (barn)	σ_i (barn)	σ_a (barn)
H	N		−3.7409(11)	1.7586(10)	79.90(4)	0.3326(7)
H	1	99.985	−3.7423(12)	1.7599(11)	79.91(4)	0.3326(7)
D	2	0.015	6.674(6)	5.597(10)	2.04(3)	0.000519(7)
T	3		4.94(8)	3.07(10)	0.00(2)	0
He	3	0.00014	5.74(7)	4.42(10)	1.2(3)	5333(7)
He	4	100	3.26(3)	1.34(2)	0	0
Li	N		−1.90(3)	0.454(14)	0.91(3)	70.5(3)
Li	6	7.5	2.0(1)	0.51(5)	0.41(11)	940(4)
Li	7	92.5	−2.22(1)	0.619(6)	0.78(3)	0.0454(3)
Be	9	100	7.79(1)	7.63(2)	0.005(1)	0.0076(8)
B	N		5.30(4)	3.54(5)	1.70(12)	767(8)
B	10	20.0	−0.1(4)	0.14(2)	3.0(4)	3837(9)
B	11	80.0	6.65(4)	5.56(7)	0.22(6)	0.0055(13)
C	N		6.6484(13)	5.554(2)	0.001(4)	0.00350(7)
N	N		9.36(2)	11.01(5)	0.49(10)	1.90(3)
O	N		5.805(4)	4.235(6)	0.000(9)	0.00019(2)
F	N		5.654(12)	4.017(17)	0.0008(2)	0.0096(5)
Ne	N		4.547(11)	2.598(13)	0.008(18)	0.039(4)
Na	N		3.63(2)	1.66(2)	1.62(3)	0.530(5)
Mg	N		5.375(4)	3.631(5)	0.077(6)	0.063(5)
Al	N		3.449(5)	1.495(14)	0.0085(7)	0.231(3)
Si	N		4.149(1)	2.163(1)	0.015(2)	0.171(3)
P	N		5.13(1)	3.307(13)	0.006(4)	0.172(6)
S	N		2.847(1)	1.0186(7)	0.007(5)	0.53(1)
Cl	N		9.5792(8)	11.531(2)	5.2(2)	33.5(3)
Ar	N		1.909(6)	0.458(3)	0.22(2)	0.675(9)

Element	A	Abundance (%)	$\bar{b}$ (fm)	σ_c (barn)	σ_i (barn)	σ_a (barn)
K	N		3.71(2)	1.73(2)	0.25(10)	2.1(1)
Ca	N		4.90(3)	3.02(4)	0.03(6)	0.43(2)
Sc	N		12.29(11)	19.0(3)	4.5(5)	27.2(2)
Ti	N		−3.30(2)	1.37(2)	2.67(4)	6.09(13)
V	N		−0.3824(12)	0.0184(1)	5.187(16)	5.08(4)
Cr	N		3.635(7)	1.660(6)	1.83(2)	3.07(8)
Mn	N		−3.73(2)	1.75(2)	0.40(2)	13.3(2)
Fe	N		9.54(6)	11.44(14)	0.39(3)	2.56(3)
Co	N		2.50(3)	0.79(2)	4.8(3)	37.18(6)
Ni	N		10.3(1)	13.3(3)	5.2(4)	4.49(16)
Ni	58	68.27	14.4(1)	26.1(4)	0	4.6(3)
Ni	60	26.10	2.8(1)	0.99(7)	0	2.9(2)
Cu	N		7.718(4)	7.486(8)	0.52(4)	3.78(2)
Zn	N		5.680(5)	4.054(7)	0.077(7)	1.11(2)
Ga	N		7.288(10)	6.67(2)	0.0(2)	2.9(1)
Ge	N		8.1929(17)	8.435(4)	0.17(6)	2.3(2)
As	N		6.58(1)	5.44(2)	0.060(9)	4.5(1)
Se	N		7.970(9)	7.98(2)	0.33(6)	11.7(2)
Br	N		6.79(2)	5.79(3)	0.10(9)	6.9(2)
Kr	N		7.80(10)	7.65(20)	0.03(24)	25(1)
Rb	N		7.08(2)	6.30(4)	0.3(2)	0.38(4)
Sr	N		7.02(2)	6.19(4)	0.04(10)	1.28(6)
Y	N		7.75(2)	7.55(4)	0.15(1)	1.28(2)
Zr	N		7.16(3)	6.44(5)	0.16(15)	0.185(3)
Nb	N		7.054(3)	6.253(5)	0.0024(3)	1.15(5)
Mo	N		6.95(7)	6.07(12)	0.28(12)	2.55(5)
Tc	99		6.8(3)	5.8(5)		20(1)
Ru	N		7.21(7)	6.53(13)	0.07(16)	2.56(13)
Rh	N		5.88(4)	4.34(6)		145(2)
Pd	N		5.91(6)	4.39(9)	0.093(9)	6.9(4)
Ag	N		5.922(7)	4.407(10)	0.58(3)	63.3(4)
Cd	N		5.1(3)	3.3(4)	2.4(7)	2520(50)
In	N		4.065(2)	2.08(2)	0.54(11)	193.8(1.5)
Sn	N		6.228(4)	4.874(6)	0.022(5)	0.626(9)
Sb	N		5.641(17)	3.999(17)	0.3(1)	5.1(1)
Te	N		5.43(4)	3.71(5)	0.02(21)	4.7(1)
I	N		5.28(2)	3.50(3)	0.0	6.2(2)
Xe	N		4.85(13)	2.96(16)		23.9(12)
Cs	N		5.42(2)	3.69(3)	0.21(5)	29.0(1,5)
Ba	N		5.25(4)	3.46(5)	0.01(6)	1.2(1)
La	N		8.24(4)	8.53(8)	1.13(19)	8.97(5)
Ce	N		4.84(2)	2.94(2)	0.00(10)	0.63(4)
Pr	N		4.45(5)	2.49(6)	0.016(3)	11.5(3)
Nd	N		7.69(5)	7.43(10)	11(2)	50.5(2.0)
Sm	N		4.2(3)	2.5(3)	50(6)	5670(100)
Eu	N		6.68(12)	5.8(2)	2.2(4)	4600(100)
Gd	N		9.5(2)	34.5(5)	158(4)	48890(104)
Tb	N		7.38(3)	6.84(6)	0.004(3)	23.4(4)
Dy	N		16.9(2)	35.9(8)	54.5(1.9)	940(15)
Ho	N		8.08(5)	8.20(10)	0.36(8)	64.7(1.2)
Er	N		8.03(3)	8.10(6)	1.2(7)	159.2(3.6)
Tm	N		7.05(5)	6.25(9)	0.41(14)	105(2)

Element	A	Abundance (%)	$\bar{b}$ (fm)	σ_c (barn)	σ_i (barn)	σ_a (barn)
Yb	N		12.40(10)	19.3(3)	3.0(5)	35.1(2.2)
Lu	N		7.3(2)	6.7(4)	0.10(5)	76.4(2.1)
Hf	N		7.77(14)	7.6(3)	2.6(5)	104.1(5)
Ta	N		6.91(7)	6.00(12)	0.020(4)	20.6(5)
W	N		4.77(5)	2.86(6)	2.00(14)	18.4(3)
Re	N		9.2(2)	10.6(5)	0.9(6)	90.7(2.1)
Os	N		11.0(2)	15.2(6)	0.4(8)	16.0(4)
Ir	N		10.6(3)	14.1(8)	0.2(2.9)	425.3(2.4)
Pt	N		9.63(5)	11.65(12)	0.13(16)	10.3(3)
Au	N		7.63(6)	7.32(12)	0.36(4)	98.65(9)
Hg	N		12.66(2)	20.14(6)	6.7(1)	372.3(4.0)
Tl	N		8.785(10)	9.70(2)	0.14(17)	3.43(6)
Pb	N		9.4003(14)	11.104(3)	0.0030(7)	0.171(2)
Bi	N		8.5256(14)	9.134(3)	0.0072(6)	0.0338(7)
Th	N		9.84(3)	12.17(7)		7.37(6)
U	N		8.417(5)	8.903(11)	0.0004(21)	7.57(2)
U	235	0.720	9.8(6)	12.1(1.5)	0.2(2)	680.5(1.3)
U	238	99.275	8.407(7)	8.882(15)	0	2.680(19)

REFERENCES

Aharony, A. (1973a). *Phys. Rev.* **B8:** 3349.
Aharony, A. (1973b). *Phys. Rev.* **B8:** 3363.
Aharony A. (1982). *Critical Phenomena* (F. J. W. Hahne, Ed.) p. 209. Berlin: Springer-Verlag.
Aharony, A. (1986). *Europhys. Lett.* **1:** 617.
Aharony, A. and Halperin, B. I. (1975). *Phys. Rev. Lett.* **35:** 1308.
Aharony, A., Imry, Y., and Ma, S-K. (1976). *Phys. Rev. Lett.* **37:** 1364.
Ahmed, N., Butt, N. M., Beg, M. M., Aslam, J., Khan, Q. H., and Collins, M. F. (1982). *Can. J. Phys.* **60:** 1323.
Aldred, A. T., du Plessis, P. de V., and Lander, G. H. (1980). *J. Magn. Magn. Mater.* **20:** 236.
Als-Nielsen, J. (1969). *Phys. Rev.* **185:** 664.
Als-Nielsen, J. (1970). *Phys. Rev. Lett.* **25:** 730.
Als-Nielsen, J. (1976a). *Phase Transitions and Critical Phenomena* (C. Domb and M. S. Green, Ed.) vol. **5A:** p. 87. New York: Academic Press.
Als-Nielsen, J. (1976b). *Phys. Rev. Let.* **37:** 1161.
Als-Nielsen, J., Axe, J. D., and Shirane, G. (1971). *J. Appl. Phys.* **42:** 1666.
Als-Nielsen, J. and Dietrich, O. W. (1967). *Phys. Rev.* **153:** 706, 711, 717.
Als-Nielsen, J., Holmes, L. M., and Guggenheim, H. J. (1974). *Phys. Rev. Lett.* **32:** 610.
Als-Nielsen, J., Holmes, L. M., Larsen, F. K., and Guggenheim, H. J. (1975). *Phys. Rev.* **B12:** 191.
Als-Nielsen, J., Birgeneau, R. J., Guggenheim, H. J., and Shirane, G. (1976a). *J. Phys.* C **9:** L121.
Als-Nielsen, J., Dietrich, O. W., and Passell, L. (1976b). *Phys. Rev.* **B14:** 4908.
Andrews, T. (1869). *Phil. Trans. R. Soc.* **159:** 575.
Arajs, S., Tehan, B. L., Anderson, E. E., and Stelmach, A. A. (1970). *Int. J. Magn.* **1:** 41.
Arrott, A., Werner, S. A., and Kendrick, H. (1965). *Phys. Rev. Lett.* **14:** 1022.
Bacon, G. E. (1975). *Neutron Diffraction,* Third Edition. Oxford: Oxford University Press.
Bak, P. and Mukamel, D. (1976). *Phys. Rev.* **B13:** 5086.
Baker, G. A. and Essam, J. W. (1971). *J. Chem. Phys.* **55:** 861.
Baker, G. A. Jr., Nickel, B. G., and Meiron, D. I. (1978). *Phys. Rev.* **B17:** 1365.
Bally, D., Popovici, M., Totia, M., Grabcev, B., and Lungu, A. M. (1968a). *Phys. Lett.* **26A:** 396.
Bally, D., Popovici, M., Totia, M., Grabcev, B., and Lungu, A. M. (1968b). *Neutron Inelastic Scattering,* Vol. II, p. 75. Vienna: IAEA.
Basten, J. A. J., Frikkee, E., and de Jonge, W. J. M. (1980). *Phys. Rev.* **B22:** 6707.
Belanger, D. P. and Yoshizawa, H. (1987). *Phys. Rev.* **B35:** 4823.
Belanger, D. P., King, A. R., and Jaccarino, V. (1984). *J. Appl. Phys.* **55:** 2383.
Belanger, D. P., King, A. R., and Jaccarino, V. (1986). *Phys. Rev.* **B34:** 452.
Berlin, T. H. and Kac, M. (1952). *Phys. Rev.* **86:** 821.
Birgeneau, R. J. (1975). In *Magnetism and Magnetic Materials—1974,* (C. D.

Graham, G. H. Lander and J. J. Rhyne, Ed.), A.I.P. Conference Proceedings No. 24, p. 258. New York: American Institute of Physics.
Birgeneau, R. J., Skalyo, J., and Shirane, G. (1971a). *Phys. Rev.* **B3**: 736.
Birgeneau, R. J., Dingle, R., Hutchings, M. T., Shirane, G., and Holt, S. L. (1971b). *Phys. Rev. Lett.* **26**: 718.
Birgeneau, R. J., Guggenheim, H. J., and Shirane, G. (1973). *Phys. Rev.* **B8**: 304.
Birgeneau, R. J., Shirane, G., Blume, M., and Koehler, W. C. (1974). *Phys. Rev. Lett.* **33**: 1098.
Birgeneau, R. J., Cowley, R. A., Shirane, G., Tarvin, J. A., and Guggenheim, H. J. (1980). *Phys. Rev.* **B21**: 317.
Birgeneau, R. J., Cowley, R. A., Shirane, G., Yoshizawa, H., Belanger, D. P., King, A. R., and Jaccarino, V. (1983a). *Phys Rev.* **B27**: 6747.
Birgeneau, R. J., Yoshizawa, H., Cowley, R. A., Shirane, G., and Ikeda, H. (1983b). *Phys. Rev.* **B28**: 1438.
Birgeneau, R. J., Shapira, Y., Shirane, G., Cowley, R. A., and Yoshizawa, H. (1986). *Physica* **137B**: 83.
Bloch, D., Hermann-Ronzaud, D., Vettier, C., Yelon, W. B., and Alben, R. (1975). *Phys. Rev. Lett.* **35**: 963.
Blume, M., Freeman, A. J., and Watson, R. E. (1962). *J. Chem. Phys.* **37**: 1245.
Blume, M., Freeman, A. J., and Watson, R. E. (1964). *J. Chem. Phys.* **41**: 1878.
Boire, R. and Collins, M. F. (1977). *Can. J. Phys.* **55**: 688.
Bongaarts, A. L. M. and de Jonge, W. J. M. (1977). *Phys. Rev.* **B15**: 3242.
Bohn, H. G., Kollmar, A., and Zinn, W. (1984). *Phys. Rev.* **B30**: 6504.
Böni, P. and Shirane G. (1986). *Phys. Rev.* **B33**: 3012.
Böni, P., Shirane, G., Bohn, H. G., and Zinn, W. (1987a). *J. Appl. Phys.* **61**: 3397.
Böni, P., Chen, M. E., and Shirane, G. (1987b). *Phys. Rev.* **B35**: 8449.
Boronkay, S. and Collins, M. F. (1973). *Int. J. Magn.* **4**: 205.
Breed, D. J. (1969). *Physics* **37**: 35.
Brockhouse, B. N. (1961). *Inelastic Scattering of Neutrons in Solids and Liquids*, p. 113. Vienna: IAEA.
Brown, P. J. (1979). *Treatise on Materials Science and Technology* (G. Kostorz, Ed.) Vol. 15. New York: Academic Press.
Buyers, W. J. L. and Holden, T. M. (1985). *Handbook on the Physics and Chemistry of the Actinides* (A. J. Freeman and G. H. Lander, Ed.) vol. 3, p. 239. Amsterdam: North-Holland.
Chipman, D. and Walker, C. (1972). *Phys. Rev.* **B5**: 3823.
Collins, M. F., Minkiewicz, V. J., Nathans, R., Passell, L., and Shirane, G. (1969). *Phys. Rev.* **179**: 417.
Coniglio, A. (1981). *Phys. Rev. Lett.* **46**: 250.
Cooper, M. J. (1968). *Phys. Rev.* **168**: 183.
Cooper, M. J. and Nathans, R. (1967). *Acta Crystallogr.* **23**: 257.
Cowley, R. A., Birgeneau, R. J., Shirane, G., Guggenheim, H. J., and Ikeda, H. (1980a). *Phys. Rev.* **B21**: 4083.
Cowley, R. A., Shirane, G., Birgeneau, R. J. Svensson, E. C., and Guggenheim, H. J. (1980b). *Phys. Rev.* **B22**: 4412.
Cowley, R. A., Hagen, M., and Belanger, D. P. (1984). *J. Phys.* **C 17**: 3763.
Dachs, H. (1978). *Neutron Diffraction* (H. Dachs, Ed.) Berlin: Springer-Verlag.
de Jongh, L. J. and Midiema, A. R. (1974). *Adv. Phys.* **23**: 1.
Dietrich, O. W. (1969). *J. Phys. C* **2**: 2022.
Dietrich, O. W. and Als-Nielsen, J. (1967). *Phys. Rev.* **162**: 315.
Dietrich, O. W. and Als-Nielsen, J. (1968). *Inelastic Neutron Scattering*, vol. II, p. 63. Vienna: IAEA.
Dietrich, O. W., Als-Nielsen, J., and Passell, L. (1976). *Phys. Rev.* **B14**: 4923.

Domb, C. (1974). *Phase Transitions and Critical Phenomena* (C. Domb and M. S. Green, Ed.) vol. 3, p. 357. New York: Academic Press.
Domb, C. and Hunter, D. L. (1965). *Proc. Phys. Soc.* **86:** 1147.
du Plessis, P. de V., van Doorn, C. F., Grobler, N. J. S., and van Delden, D. C. (1982). *J. Phys. C.* **15:** 1525.
du Plessis, P. de V., van Doorn, C. F., and van Delden, D. C. (1983). *J. Magn. Magn. Mater.* **40:** 91.
Eckeret, J. and Shirane, G. (1976). *Solid State Commun.* **19:** 911.
Enz, C. P. (Editor) (1979). *Dynamic Critical Phenomena and Related Topics.* Berlin: Springer-Verlag.
Essam, J. W. (1980). *Rep. Prog. Phys.* **43:** 833.
Ferer, M. and Hamid-Aidinejad, A. (1986). *Phys. Rev.* **B34:** 6481.
Ferrell, R. A., Menyhard, N., Schmidt, H., Schwabb, F., and Szepfalusy, P. (1968). *Ann. Phys. (N.Y.)* **47:** 565.
Fincher Jr., C. R., Shirane, G., and Werner, S. A. (1981). *Phys. Rev.* **B24:** 1312.
Finger, W. (1977). *Phys. Lett.* **60A;** 165.
Fisher, M. E. (1964). *J. Math. Phys.* **5:** 944.
Fisher, M. E. (1967). *Rep. Prog. Phys.* **30:** 615.
Fisher, M. E. (1974). *Magnetism and Magnetic Materials*—1974, (C. D. Graham, G. H. Lander and J. J. Rhyne, Ed.) A.I.P. Conference Proceedings No. 24, p. 273. New York: American Institute of Physics.
Fisher, M. E. (1982). *Critical Phenomena* (F. J. W. Hahne, Ed.) Lecture Notes in Physics #186, p. 1–139. Berlin: Springer-Verlag.
Fisher, M. E. and Aharony, A. (1973). *Phys. Rev. Lett.* **30:** 559.
Fisher, M. E. and Burford, R. J. (1967). *Phys. Rev.* **156:** 583.
Fishman, S. and Aharony, A. (1979). *J. Phys. C* **12:** L729.
Fitzgerald, W. J., Visser, D., and Ziebeck, K. R. A. (1982). *J. Phys. C* **15:** 795.
Fixman, M. (1962). *J. Chem. Phys.* **36:** 310.
Folk, R. and Iro, H. (1985). *Phys. Rev.* **B32:** 1880.
Forster, D. (1975). *Hydrodynamic Fluctuations, Broken Symmetry, and Correlation Functions.* Reading, Mass.: W. A. Benjamin,
Gaulin, B. D. and Collins, M. F. (1984). *Can J. Phys.* **62:** 1132.
George, M. J. and Rehr, J. J. (1984). *Phys. Rev. Lett* **53:** 2063.
Glinka, C. J., Minkiewicz, V. J., and Passell, L. (1977). *Phys. Rev.* **B16:** 4084.
Goldstone, J., Salam, A., and Weinberg, S. (1962). *Phys. Rev.* **127:** 965.
Greissler, K. K. and Lange, H. (1966). *Z. Angew. Phys.* **21:** 357.
Griffith, R. B. (1965). *J. Chem. Phys.* **43:** 1958.
Griffith, R. B. (1970). *Phys. Rev. Lett.* **24:** 1479.
Griffith, R. B. and Wheeler, J. C. (1970). *Phys. Rev.* **A2:** 1047.
Grinstein, G. (1984). *J. Appl. Phys.* **55:** 2371.
Guttmann, A. J. (1978). *J. Phys. A* **11:** 545.
Guttman, L. and Schynders, H. C. (1969). *Phys. Rev. Let.* **22:** 520.
Guttman, L., Schynders, H. C., and Arai, G. J. (1969). *Phys. Rev. Let.* **22:** 517.
Hagen, M. and Paul D. McK. (1984). *J. Phys. C* **17:** 5605.
Hagen, M., Cowley, R. A., Nicklow, R. M., and Ikeda, H. (1987). *Phys. Rev.* **B36:** 401.
Hagen, M., Stirling, W. G., and Lander, G. H. (1988). *Phys. Rev.* **B37:** 1846.
Haldane, F. D. M. (1983). *Phys. Rev. Lett.* **50:** 1153; *Phys. Lett.* **93A:** 464.
Halg, B., Furrer, A., Halg, W., and Vogt, O. (1982a). *J. Magn. Magn. Mater.* **29:** 151.
Halg, B., Furrer, A., Halg, W., and Vogt, O. (1982b). *J. Appl. Phys.* **53:** 1927.
Halperin, B. I. and Hohenberg, P. C. (1969a). *Phys. Rev.* **177:** 952.
Halperin, B. I. and Hohenberg, P. C. (1969b). *Phys. Rev.* **188:** 898.
Halperin, B. I. and Hohenberg, P. C. (1977). *Rev. Mod. Phys.* **49:** 435.
Halperin, B. I., Hohenberg, P. C., and Ma, S. (1972). *Phys. Rev. Lett.* **29:** 1548.

Halperin, B. I., Hohenberg, P. C., and Ma. S. (1974). *Phys. Rev.* **B10**: 139.
Hamaguchi, Y., Tsunoda, Y., and Kunitomi, N. (1968). *J. Appl. Phys.* **39**: 1227.
Harris, A. B. (1974). *J. Phys. C* **7**: 1671.
Heller, P. (1966). *Phys. Rev.* **146**: 403.
Heller, P. and Benedek, G. B. (1962). *Phys. Rev. Let.* **8**: 428.
Hirakawa, K. (1982). *J. Appl. Phys.* **53**: 1893.
Hirakawa, K. and Ikeda, H. (1973). *J. Phys. Soc. Japan* **35**: 1328.
Hirakawa, K. Yoshizawa, H. and Ubukoshi, K. (1982). *J. Phys. Soc. Japan* **51**: 2151.
Hirakawa, K. Yoshizawa, H., Axe, J. D., and Shirane, G. (1983). *J. Phys. Soc. Japan* **52**: 4220.
Hirte, J., Weitzel, H. and Lehner, N. (1984). *Phys. Rev.* **B30**: 6707.
Hohenemser, C., Chow, L., and Suter, R. M. (1982). *Phys. Rev.* **B26**: 5056.
Holden, T. M., Buyers, W. J. L., Svensson, E. C., and Lander, G. H. (1982). *Phys. Rev.* **B26**: 6227.
Holstein, T. and Primakoff, H. (1940). *Phys. Rev.* **58**: 1098.
Hornreich, R. M. (1980). *J. Mag. Magn. Mater.* **15–18**: 387.
Hu, B. (1982). *Phys. Rep.* **91**: 233.
Hubbard, J. (1971). *J. Phys. C* **4**: 53.
Hughes, D. J. and Palevsky, H. (1953). *Phys. Rev.* **92**: 202.
Hughes, K. M., Holden, T. M., Buyers, W. J. L., Collins, M. F., and du Plessis, P. de V. (1987). *J. Appl. Phys.* **61**: 3412.
Hutchings, M. T., Schulhof, M. P. and Guggenheim, H. J. (1972). *Phys. Rev.* **B5**: 154.
Hutchings, M. T., Day, P., Janke, E. and Pynn, R. (1986). *J. Magn. Magn. Mater.* **54-57**: 673.
Ikeda, H. and Hirakawa, K. (1972). *J. Phys. Soc. Japan* **33**: 393.
Ikeda, H. and Hirakawa, K. (1974). *Solid State Commun.* **14**: 529.
Ikeda, H. and Hutchings, M. T. (1978). *J. Phys. C* **11**: L529.
Ikeda, H., Suzuki, M. and Hutchings, M. T. (1979). *J. Phys. Soc. Japan* **46**: 1153.
Imry, Y. and Ma, S-K. (1975). *Phys. Rev. Lett.* **35**: 1399.
Iyengar, P. K. (1965). *Thermal Neutron Scattering* (P.A. Egelstaff, Ed.) New York: Academic Press.
Ishikawa, Y., Noda, Y., Fincher, C. R., and Shirane, G. (1982). *Phys. Rev.* **B25**: 254.
Ishikawa, Y., Uemura, Y. J., Majkrzak, C. F., Shirane, G., and Noda, Y. (1985). *Phys. Rev.* **B31**: 5884.
Jug, G. (1983). *Phys. Rev.* **B27**: 609.
Kadanoff, L. P. (1971). *Proc. 1970 Varenna Summer School on Critical Phenomena* (M. S. Green, Ed.), pp. 100–117. New York: Academic Press.
Kadanoff, L. P. and Swift, J. (1968). *Phys. Rev.* **166**: 89.
Kadowaki, H., Ubukoski, K., Hirakawa, K., Martinez, J. L., and Shirane, G. (1987). *J. Phys. Soc. Japan* **56**: 1294.
Kawamura, H. (1985). *J. Phys. Soc. Japan* **54**: 3220.
Kawamura, H. (1986). *J. Phys. Soc. Japan* **55**: 2095.
Kawasaki, K. (1967). *J. Phys. Chem. Solids* **28**: 1277.
Kawasaki, K. (1970). *Ann. Phys. (N.Y.)* **61**: 1.
Kawasaki, K. (1976). *Phase Transitions and Critical Phenomena*, (C. Domb and M. S. Green, Ed.) Volume 5A. New York: Academic Press.
Kopinga, K., Steiner, M. and de Jonge, W. J. M. (1985). *J. Phys. C* **18**: 3511.
Kosterlitz, J. M. (1974). *J. Phys. C* **7**: 1046.
Kosterlitz, J. M. and Thouless, D. J. (1973). *J. Phys. C* **6**: 1181.
Kötzler, J. (1983). *Phys. Rev. Lett.* **51**: 833.
Kötzler, J., Gorlitz, D., Mezei, F., and Farago, B. (1986). *Europhys. Lett.* **1**: 675.

Kouvel, J. S., Kouvel, J. B., and Comly, J. B. (1968). *Phys. Rev. Lett.* **20:** 1237.
Kuznietz, M., Burlet, P., Rossat-Mignod, J., and Vogt, O. (1985). *J. Magn. Magn. Mater.* **54–57:** 553.
Landau, L. D. and Lifshitz, E. M. (1969). *Statistical Physics*, Second Edition. Reading, Mass: Addison-Wesley.
Lander, G. H., Sinha, S. K., Sparlin, D. M., and Vogt, O. (1978). *Phys. Rev. Lett.* **40:** 523.
Larkin, A. I. and Khmel'nitzkii (1969). *Zh. Eksp. Teor. Fiz.* **56:** 2087. (*Sov. Phys. JETP* **29:** 1123.)
Le Guillou, J. C. and Zinn-Justin, J. (1980). *Phys. Rev.* **B21:** 3976.
Lindgard, P. A. (1978). In *Neutron Diffraction* (H. Dachs, Ed.), p. 197. Berlin: Springer-Verlag.
Loopstra, B. O. (1966). *Nucl. Instrum. Meth.* **44:** 181.
Lovesey, S. W. (1984). *Theory of Neutron Scattering from Condensed Matter.* Oxford: Oxford University Press.
Lovesey, S. W. and Williams, R. D. (1986). *J. Phys. C* **19:** L253.
Lubensky, T. C. (1977). *Phys. Rev.* **B15:** 311.
Lynn, J. W. and Mook, H. A. (1986). *J. Magn. Magn. Mater.* **54–57:** 1169.
Lyons, D. H. and Kaplan, T. A. (1960). *Phys. Rev.* **120:** 1580.
Martinez, J. L., Böni, P., and Shirane, G. (1985). *Phys. Rev.* **B32:** 7037.
Mason, T. E., Collins, M. F. and Gaulin, B. D. (1987). *J. Phys. C:* **20:** L947.
Mermin, N. D. and Wagner, H. (1966). *Phys. Rev. Lett.* **17:** 1133.
Mezei, F. (1980a). *Imaging Processes and Coherence in Physics* (M. Schlenker, Ed.) p. 282. Berlin: Springer-Verlag.
Mezei, F. (Editor) (1980b). *Neutron Spin Echo*, p. 1. Berlin: Springer-Verlag.
Mezei, F. (1984). *J. Magn. Magn. Mater.* **45:** 67.
Mezei, F. (1986). *Physica* **136B:** 417.
Michelson, A. (1977). *Phys. Rev.* **B16:** 577, 585, 5121.
Minkiewicz, V. J., Collins, M. F., Nathans, R., and Shirane, G. (1969). *Phys. Rev.* **182:** 624.
Minkiewicz, V. J., Gesi, K., and Hirahara, E. (1971). *J. Appl. Phys.* **42:** 1374.
Mitchell, P. W. and Paul, D. McK. (1986). *J. Magn. Magn. Mater.* **54–57:** 1154.
Mitchell, P. W., Cowley, R. A., Yoshizawa, H., Böni, P., Uemura, Y. J., and Birgeneau, R. J. (1986). *Phys. Rev.* **B34:** 4719.
Moller, J. B., Lander, G. H., and Vogt, O. (1979). *J. de Physique* **C4:** 28.
Moon, R. M., Cable, J. W., and Shapira, Y. (1981) *J. Appl. Phys.* **52:** 2025.
Moriya, T. (1985). *Spin Fluctuations in Itinerant Electron Magnetism.* Berlin: Springer-Verlag.
Moussa, F., Hennion, B., Mons, J., and Pepy, G. (1978). *Solid State Commun.* **27:** 141.
Mukamel, D. (1977). *J. Phys.* **A10:** L249.
Mukamel, D. and Korinsky, S. (1976). *Phys. Rev.* **B13,** 5065.
Negovetic, I. and Konstantinovic, J. (1973). *Solid State Commun.* **13:** 249.
Neimeijer, Th. and van Leeuwen, J. M. J. (1973). *Phys. Rev. Lett.* **31:** 1411.
Neimeijer, Th. and van Leeuwen, J. M. J. (1974). *Physica* **71:** 17.
Nielson, M. and Moller, H. B. (1969). *Acta Crystallogr.* **A25:** 567.
Noakes, J. E., Tornberg, N. E., and Arrott, A. (1966). *J. Appl. Phys.* **37:** 1264.
Noda, Y. and Ishikawa, Y. (1976). *J. Phys. Soc. Japan* **40:** 690, 699.
Norvell, J. C., Wolf, W. P., Corliss, L. M., Hastings, J. M., and Nathans, R. (1969a). *Phys. Rev.* **186:** 557.
Norvell, J. C., Wolf, W. P., Corliss, L. M., Hastings, J. M., and Nathans, R. (1969b). *Phys. Rev.* **186:** 567.
Norvell, J. C. and Als-Nielsen, J. (1970). *Phys. Rev.* **B2:** 277.
Onsager, L. (1944). *Phys. Rev.* **65:** 117.
Ornstein, L. S. and Zernicke, F. (1914). *Proc. Sect. Sci. K. Med. Akad, Wet.* **17:** 793.

Ott, H. R., Kjems, J. K., and Hulliger, F. (1979). *Phys. Rev. Lett.* **42:** 1378.
Oyedele, J. A. and Collins, M. F. (1977). *Phys. Rev.* **B16:** 3208.
Parette, G. and Kahn, R. (1971). *J. de Physique* **32:** 447.
Passell, L., Als-Nielsen, J., and Dietrich, O. W. (1972). *Neutron Inelastic Scattering* 1972, p. 619. Vienna: IAEA.
Passell, L., Dietrich, O. W., and Als-Nielsen, J. (1976). *Phys. Rev.* **B14:** 4897.
Pelcovits, R. A. and Aharony, A. (1985). *Phys. Rev.* **B31:** 350.
Pfeuty, P. and Toulouse, G. (1977). *Introduction to Renormalization Groups and Critical Phenomena.* New York: Wiley.
Pfeuty, P., Jassnow, D., and Fisher, M. E. (1974). *Phys. Rev.* **B10:** 2088.
Praveczki, E. (1980). *J. Phys. C.* **13:** 2161.
Pynn, R. and Skjeltorp, A. (Editors) (1984). *Multicritical Phenomena.* New York: Plenum Press.
Rácz, Z. and Collins, M. F. (1980). *Phys. Rev.* **B21:** 229.
Rathmann, O. and Als-Nielsen, J. (1974). *Phys. Rev.* **9:** 3921.
Regnault, L. P., Rossat-Mignod, J., and Henry, J. Y. (1983). *J. Phys. Soc. Japan* **52:** Suppl. 1.
Résibois, P. and Piette, C. (1970). *Phys. Rev. Lett.* **24:** 514.
Riedel, E. and Wegner, F. (1969). *Z. Physik* **225:** 195.
Riedel, E. and Wegner, F. (1970). *Phys. Rev. Lett.* **24:** 730; *Phys. Rev. Lett.* **24:** 930E; *Phys. Lett.* **32A:** 273.
Ritchie, D. S. and Fisher, M. E. (1972). *Phys. Rev.* **B5:** 2668.
Rocker, W. and Kohlhass, R. (1967). *Z. Naturforsch.* **22A:** 291.
Rossat-Mignod, J., Lander, G. H., and Burlet, P. (1984). *Handbook on the Physics and Chemistry of the Actinides* (A. J. Freeman and G. H. Lander, Ed.) vol. 1, p. 415. Amsterdam: North-Holland.
Rushbrooke, G. S. (1963). *J. Chem. Phys.* **39:** 842.
Samuelsen, E. J. (1973). *Phys. Rev. Lett.* **31:** 936.
Schobinger-Papamantellos, P., Fisher, P., Niggli, A., Kaldis, E., and Hildebrandt, V. (1974). *J. Phys. C.* **7:** 2023.
Schulhof, M. P., Heller, P., Nathans, R., and Linz, A. (1970). *Phys. Rev.* **B1:** 2304.
Schulhof, M. P., Nathans, R., Heller, P., and Linz, A. (1971). *Phys. Rev.* **B4:** 2254.
Sears, V. F. (1984). AECL report AECL-8490 and to appear in the *International Tables for Crystallography,* Volume C (Theo. Hahn, Ed.) Dordrecht: D. Reidel.
Shapira, Y. (1984). In *Multicritical Phenomena* (R. Pynn and A. Skjeltorp, Ed.) p. 53. New York: Plenum Press.
Shapiro, S. M. and Bak, P. (1975). *J. Phys. Chem. Solids* **36:** 579.
Shenker, S. H. and Tobochnik, J. (1980). *Phys. Rev.* **B22:** 4462.
Shirane, G. and Minkiewicz, V. J. (1970). *Nucl. Instrum. Meth.* **89:** 109.
Shirane, G., Uemura, Y. J., Wickstead, J. P., Endoh, Y., and Ishikawa, Y. (1980). *Phys. Rev.* **B31:** 1227.
Shull, C. G. (1967). *Trans. Amer. Crystallogr. Ass.* **3:** 1.
Sinha, S. K., Lander, G. H., Shapiro, S. M., and Vogt, O. (1981). *Phys. Rev.* **B23:** 4556.
Squires, G. L. (1978). *Introduction to the Theory of Thermal Neutron Scattering,* Cambridge University Press.
Squires, G. L. (1954). *Proc. Phys. Soc.* **67:** 248.
Stanley, H. E. (1968). *Phys. Rev.* **176:** 718.
Stanley, H. E. (1971). *Introduction to Phase Transitions and Critical Phenomena.* Oxford University Press.
Stanley, H. E. and Kaplan, T. A. (1966). *Phys. Rev. Lett.* **17:** 913.

Stanley, H. E., Birgeneau, R. J., Reynolds, P. J., and Nicoll, J. E. (1976). *J. Phys. C* **9**: L553.
Stauffer, D. (1976). *Z. Phys.* **B22**: 161.
Steiner, M. and Dachs, H. (1971). *Solid State Commun.* **9**: 1603.
Steiner, M., Villain, J., and Windsor, C. G. (1976). *Adv. Phys.* **25**: 87.
Stinchcombe, R. B. (1983). *Phase Transitions and Critical Phenomena* (C. Domb and J. L. Lebowitz, eds.) vol. 7, p. 152. New York: Academic Press.
Stüsser, N., Rekveldt, M. Th., and Spruijt, T. (1986). *J. Magn. Magn. Mater.* **54–57**: 723.
Tarko, H. B. and Fisher, M. E. (1975). *Phys. Rev.* **B11**: 1217.
Thorpe, M. F. (1975). *J. de Physique* **36**: 1177.
Tillwick, D. L. and du Plessis, P. de V. (1976). *J. Magn. Magn. Mater.* **3**: 319.
Tracy, C. A. and McCoy, B. (1975). *Phys. Rev.* **B12**: 368.
Tucciarone, A., Lau, H. Y., Corliss, L. M., Delapalme, A., and Hastings, J. M. (1971). *Phys. Rev.* **B4**: 3206.
Van Hove, L. (1954a). *Phys. Rev.* **95**: 249.
Van Hove, L. (1954b). *Phys. Rev.* **95**: 1374.
Wallace, D. J. and Young, A. P. (1978). *Phys. Rev.* **B17**: 2384.
Watson, R. E. and Freeman, A. J. (1961). *Acta Crystallogr.* **14**: 27.
Werner, S. A., Arrott, A., and Kendrick, H. (1967). *Phys. Rev.* **115**: 528.
Wertheim, G. K. (1967). *J. Appl. Phys.* **38**: 971.
Widom, B. (1965). *J. Chem. Phys.* **43**: 3898.
Wilson, K. G. (1971). *Phys. Rev.* **B4**: 3174, 3189.
Wilson, K. G. and Kogut, J. (1974). *Phys. Rep.* **12**: 75.
Wilson, K. G. and Fisher, M. E. (1972). *Phys. Rev. Lett.* **28**: 240.
Windsor, C. G. (1978). *Physica* **91B**: 119.
Wong, P., Cable, J. W., and Dimon, P. (1984). *J. Appl. Phys.* **55**: 2377.
Wu, T. T. (1966). *Phys. Rev.* **149**: 380.
Yamada, K., Todate, Y., Endoh, Y., Ishikawa, Y., Böni, P., and Shirane, G. (1987). *J. Appl. Phys.* **61**: 3400.
Yoshizawa, H., Cowley, R. A., Shirane, G., Birgeneau, R. J., Guggenheim, H. J., and Ikeda, H. (1982). *Phys. Rev. Lett.* **48**: 438.

AUTHOR INDEX

Aharony, A. 52, 57, 61, 111, 116, 117, 121, 128, 168
Ahmed, N. 118
Alben, R. 162
Aldred, A. T. 158
Als-Nielsen, J. 89, 93, 106–7, 116–126, 134, 148, 152, 156
Anderson, E. E. 143
Andrews, T. 3, 7
Arai, G. J. 118
Arajs, S. 143
Arrott, A. S. 12–13, 143, 150–1
Aslam, J. 118
Axe, J. D. 105, 152

Bacon, G. E. 73, 84, 89
Baker, G. A. Jr 29, 41, 118, 120
Bak, P. 18, 156
Bally, D. 141–3
Basten, J. A. J. 136–9
Beg, M. M. 118
Belanger, D. P. 97–100, 112, 165–6, 169
Benedek, G. B. 112
Berlin, T. H. 18, 19
Birgeneau, R. J. 30, 60, 97, 105–9, 134, 136, 164–9
Bloch, D. 162
Blume, M. 72, 136
Bohn, H. G. 122, 127
Boire, R. 162
Bongaarts, A. L. M. 135–6
Böni, P. 122, 125–7, 140, 154–5, 164–5
Boronkay, S. 145–8
Breed, D. J. 95
Brockhouse, B. N. 90
Brown, P. J. 89
Burford, R. J. 26–8, 97, 112, 120
Burlet, P. 159, 161, 163
Butt, N. M. 118
Buyers, W. J. L. 157–161

Cable, J. W. 138–9, 169
Chen, M. E. 125–6
Chipman, D. 118–9
Chow, L. 147
Collins, M. F. 110, 118, 120, 134, 141–9, 157, 162–3
Comly, J. B. 143

Coniglio, A. 167
Cooper, M. J. 21, 88
Corliss, L. M. 112, 128–132, 134
Cowley, R. A. 30, 60, 97–100, 164–9

Dachs, H. 89, 109
Day, P. 105
de Jonge, W. J. M. 109, 135–9
de Jongh, L. J. 110, 134
Delapalme, A. 128–132, 134
Dietrich, O. W. 118, 120–126, 128–132, 134, 156
Dimon, P. 169
Dingle, R. 108–9
Domb, C. 21, 118
du Plessis, P. deV. 156–8

Eckert, J. 156
Ehrenfest, P. 4
Endoh, Y. 154–5
Enz, C. P. 47
Essam, J. W. 60, 118, 120

Farago, B. 122
Ferer, M. 29
Ferrell, R. A. 42
Fincher, C. R. Jr 150–3
Finger, W. 47
Fisher, M. E. 15, 26–8, 30, 38, 41, 49, 51, 57, 97, 109, 112, 120–1
Fisher, P. 161
Fishman, S. 61
Fitzgerald, W. J. 109
Fixman, M. 45
Folk, R. 47, 126
Forster, D. 47
Freeman, A. J. 72
Frikkee, E. 136–9
Furrer, A. 158, 161

Gaulin, B. D. 110, 134
George, M. J. 29, 165
Gesi, K. 154–5
Glinka, C. J. 141–9, 163
Goldstone, J. 6
Gorlitz, D. 122

Author Index

Grabcev, B. 141–3
Greissler, K. K. 143
Griffiths, R. B. 15–16
Grinstein, G. 61, 62
Grobler, N. J. S. 158
Guggenheim, H. J. 30, 60, 97, 106–7, 112–5, 117, 134, 166–7, 169
Gunton, J. D. 47
Guttmann, A. J. 106
Guttman, L. 118, 120

Hagen, M. 96–100, 158–160, 164, 166
Haldane, F. D. M. 17
Halg, B. 158, 161
Halg, W. 158, 161
Halperin, B. I. 17, 42, 45–47, 117, 124
Hamaguchi, Y. 150
Hamid-Aidinejad, A. 29
Harris, A. B. 58
Hastings, J. M. 112, 128–132, 134
Heller, P. 112–3, 115, 165
Hennion, B. 100, 105
Henry, J. Y. 103–4
Hermann-Ronzaud, D. 162
Hildebrandt, V. 161
Hirahara, E. 154–5
Hirakawa, K. 96, 100–103, 105, 134
Hirte, J. 136
Hohenberg, P. C. 17, 42, 45–47, 124
Hohenemser, C. 147
Holden, T. M. 157–161
Holmes, L. M. 117
Holstein, T. 123
Holt, S. L. 108–9
Hornreich, R. M. 80
Hubbard, J. 47
Hughes, D. J. 94
Hughes, K. M. 157
Hulliger, F. 161–2
Hunter, D. L. 21
Hutchings, M. T. 95–96, 100, 105, 108–9, 112–5, 164
Hu, B. 32, 41

Ikeda, H. 95–96, 100–103, 105, 164, 166–9
Imry, Y. 61
Iro, H. 47, 126
Ishikawa, Y. 153–5
Iyengar, P. K. 85, 89

Jaccarino, V. 165, 169
Janke, E. 105
Jassnow, D. 49, 51
Jug, G. 59, 165

Kac, M. 18, 19
Kadanoff, L. P. 16, 19–20, 30, 45

Kadowaki, H. 134
Kahn, R. 148–9
Kaldis, E. 161
Kaplan, T. A. 56, 106–7
Kawamura, H. 18, 132–4
Kawasaki, K. 45
Kendrick, H. 150–1
Khan, Q. H. 118
Khmel'nitzkii, D. E. 116
King, A. R. 165, 169
Kjems, J. K. 161–2
Koehler, W. C. 136
Kogut, J. 31, 36, 41
Kohlhass, R. 143
Kollmar, A. 122, 127
Konstantinovic, J. 163
Kopinga, K. 109
Kosterlitz, J. M. 29, 94, 101–3
Kötzler, J. 122, 126–7, 147
Kouvel, J. S. 143
Krinsky, S. 18, 160, 162
Kunitomi, N. 150
Kuznietz, M. 161

Landau, L. D. 163
Lander, G. H. 158–163
Lange, H. 143
Larkin, A. I. 116
Larsen, F. K. 117
Lau, H. J. 128–132, 134
Le Guillou, J. C. 41
Lehner, N. 136
Lifshitz, E. M. 163
Lingard, P. A. 139
Linz, A. 112–3, 115, 165
Loopstra, B. O. 85
Lovesey, S. W. 66–70, 73, 80, 83–84, 126, 147
Lubensky, T. C. 60
Lungu, A. M. 141–3
Lynn, J. W. 140
Lyons, D. H. 56

Majkrzak, C. F. 153
Martinez, J. L. 134, 140
Mason, T. E. 134
Mazendko, G. F. 47
Ma, S. 17, 61
Meiron, D. I. 29, 41
Menyhard, N. 42
Mermin, N. D. 106
Mezei, F. 91–3, 122, 125, 147–9
Michelson, A. 57
Midiema, A. R. 110, 134
Minkiewicz, V. J. 85, 141–9, 154–5, 163
Mitchell, P. W. 141, 164–5, 169
Moller, H. B. 88, 158
Mons, J. 100, 105

Mook, H. A. 140
Moon, R. M. 138–9
Moriya, T. 140, 153, 163
Moussa, F. 100, 105
Mukamel, D. 18, 139, 156, 160, 162

Nathans, R. 88, 112–3, 115, 141–9, 163, 165
Negovetic, I. 163
Neimeijer, Th. 32
Nickel, B. G. 29, 41
Nicklow, R. M. 164
Nicoll, J. E. 60
Nielson, M. 88
Niggli, A. 161
Noakes, J. E. 12–13, 143
Noda, Y. 153, 155
Norvell, J. C. 112, 118

Onsager, L. 19, 94
Ornstein, L. S. 26, 28, 94
Ott, H. R. 161–2
Oyedele, J. A. 118

Palevsky, H. 94
Parette, G. 148–9
Passell, L. 121–6, 141–9, 163
Paul, D. McK. 96, 99–100, 141
Pelcovits, R. A. 168
Pepy, G. 100, 105
Pfeuty, P. 41, 49, 51
Piette, C. 47, 148, 153–5
Popovici, M. 141–3
Pravecki, E. 106
Primakoff, H. 123
Pynn, R. 139

Rácz, Z. 118, 120
Rathmann, O. 118–9
Regnault, L. P. 103–4
Rehr, J. J. 29, 165
Rekveldt, M. Th. 143
Résibois, P. 47, 148, 153–5
Reynolds, P. J. 60
Riedel, E. 47, 49, 114, 122
Ritchie, D. S. 27
Rocker, W. 143
Rossat-Mignod, J. 103–4, 159, 161, 163
Rushbrooke, G. S. 15

Salam, A. 6
Samuelsen, E. J. 96, 100
Schmidt, H. 42
Schnyders, H. C. 118, 120
Schobinger-Papamantellos, P. 161
Schulhof, M. P. 112–5, 165
Schwabb, F. 42

Sears, V. F. 170–2
Shapira, Y. 137–9, 169
Shapiro, S. M. 18, 161–2
Shenker, S. H. 106
Shirane, G. 30, 60, 85, 97, 105–9, 122, 125–7, 134, 136, 140–156, 163, 165–9
Shull, C. G. 64
Sinha, S. K. 158–162
Skalyo, J. 105–7
Skjeltorp, A. 139
Sparlin, D. M. 158–160
Spruijt, T. 143
Squires, G. L. 66–70, 73, 80, 83–4, 94
Stanley, H. E. 7, 11, 15, 18–19, 30, 47, 60, 106–7
Stauffer, D. 60
Steiner, M. 109–110
Stelmach, A. A. 143
Stinchcombe, R. B. 59–60, 62
Stirling, W. G. 158–160
Stüsser, N. 143
Suter, R. M. 147
Suzuki, M. 47, 96, 100, 164
Svensson, E. C. 158–161, 167
Swift, J. 45
Szepfalusy, P. 42

Tarko, H. B. 97, 112
Tarvin, J. A. 30, 60, 167, 169
Tehan, B. L. 143
Thorpe, M. F. 60
Thouless, D. J. 94
Tillwick, D. L. 158
Tobochnik, J. 106
Todate, Y. 154–5
Tornberg, N. E. 12–13, 143
Totia, M. 141–3
Tsunoda, Y. 150
Tucciarone, A. 128–132, 134

Ubukoshi, K. 103, 134
Uemura, Y. J. 153, 155, 164–5, 169

van Delden, D. C. 156, 158
Vogt, O. 158–162

Wagner, H. 106
Walker, C. 118–9
Wallace, D. J. 60
Watson, R. E. 72
Wegner, F. 47, 49, 114, 122
Weinberg, S. 6
Weitzel, H. 136
Werner, S. A. 150–2
Wertheim, G. K. 112

Wheeler, J. C. 16
Wicksted, J. 155
Widom, B. 21
Williams, R. D. 126, 147
Wilson, K. G. 31–32, 36, 38, 41
Windsor, C. G. 110, 140
Wolf, W. P. 112
Wong, P. 169
Wu, T. T. 97

Yamada, K. 154–5
Yelon, W. B. 162
Yoshizawa, H. 103, 105, 112, 164–5, 168–9
Young, A. P. 60

Ziebeck, K. R. A. 109
Zinn, W. 122, 127
Zinn-Justin, J. 41

SUBJECT INDEX

Absorption cross section 66, 170–2
Actinides 140–1, 156–163
Alloys, ordering 18, 117–120
Analyzer, 85, 90
Anisotropic effects 48, 105–9, 111, 113, 133, 140, 157–160
Asymptotic effects 48
Atomic form factor 72

$BaM_2(XO_4)_2$ 103–4
Beryllium filter 85
Beta brass 117–120
Bicritical point 54–55, 136–8
Borderline dimension 37, 116–7
Born approximation 66
Bragg angle 82
Bragg scattering of neutrons 79–88, 128, 149–151, 153, 160
Bragg—Williams theory 8
Bragg's Law 82, 87
Brass, beta 117–120
Broken symmetry 6, 60–61

Carbon dioxide 3
CeAs 161
CeBi 161
Central mode 113, 123, 131–2, 154–5
Cerium compounds 140–1, 156, 161–3
CeS 161
CeSb 161
CeSe 161–3
CeTe 161–3
Characteristic frequency 43, 78–9, 114–5, 124–8, 146–8, 154
Chromium 140, 149–152, 156, 161, 163
Cobalt, 125, 140–9, 153–4
Coherent scattering cross section 68–9
Coherent scattering of neutrons 68–9
Cold source of neutrons 63
Collimation of neutron beams 85–8
Compressible Ising model 118–120
Conservation laws 17
Constant energy scan 90
Constant Q scan 90
Correction terms 14, 53, 116
Correlation function, dynamic 43–6, 74–7, 113–6, 122–8, 131

Correlation function, static 25–8, 77–80, 97–9, 101–3, 120, 128–130, 166–8
Correlated region 20, 94
Correlation length 6, 13, 21, 60, 94, 97–9, 101, 113, 116–7, 121, 151, 160–1, 166–8
Corresponding states, law of 16
Critical dynamics 42–7, 90–3, 122–8, 130–2, 143–9, 151–4
Critical exponent 12–15, 20–30, 36, 111–169
Critical exponents predicted values from models 29, 35–6, 41, 45–6, 101, 118, 121
Critical point 3, 35
Critical slowing down 6
Cross over behavior 48–51, 55, 95, 100–1, 105–6, 111–3, 117, 121–2, 125, 133–4, 147, 155, 160, 164
Cross over exponents 48–51, 61
Cross section 65–6, 170–2
Cross section, magnetic 69–72, 76–80
Cross section, nuclear 67–9, 170–2
Crystal field effects 111, 140, 162
Crystal rotation method 83–4, 128
$CsCoBr_3$ 94
$CsCoCl_3.2D_2O$ 135–6
$CsMnBr_3$ 94, 133–4
$CsMnBr_3.2D_2O$ 136
$CsNiF_3$ 94
Curie constant 11
Curie-Weiss law 11
Curie temperature 5
CuZn 117–120

DAG 111–112
Damped spin waves 105, 107, 113, 123, 130–2, 143–5, 157
Debye-Waller factor 83
Depolarization of neutron beams 143
Detailed balance 43
Differential cross section 65, 78–9, 83
Diffuse scattering of neutrons 80
Dilution 58–9, 61, 164–6
Dimensionality of order-parameter 6, 16–19, 156
Dimensionality of system 16–19
Dipolar interactions 111, 116–7, 121–2, 125–8, 141, 147
Double K structure 161
Dynamic critical exponents 17, 114–6, 122–5

185

Subject Index

Dynamic scaling 42–7, 77, 114–5, 124–6, 129, 146–9, 153–6
Dynamics, critical 42–7, 105, 113–6, 122–8, 130–2, 143–9, 151–4
Dysprosium 140, 156
Dysprosium aluminium garnet 111–3

Easy plane 48
Elastic scattering of neutrons 79–80, 95
Erbium, form factor 72
Energy gap 154
Epsilon expansion 38–41
EuO 121–8, 143–5, 147
EuS 121–8, 147
Ewald sphere 83

Fan phase 139
Fe 117, 140–9
Fe_3Al, 118, 120
$FeCl_2$ 136
FeCo 118
FeF_2 111–6, 164–5
Fermi pseudopotential 67
Ferroelectricity 11
$Fe_xZn_{1-x}F_2$ 164–5, 169
Filter, neutron 85
First Born approximation 66
First order phase transition 3–4, 52, 135, 150–1, 161–3
Fisher and Burford correlation function 27–8, 97, 112, 120
Fixed point 31, 35, 38, 162
Focusing effects in neutron beams 88
Form factor, atomic 72
Form factor, magnetic 71–2
Fourier transform 27, 32, 43, 76, 79, 92, 97–8, 125
Frustration 157

Gadolinium 140
Gaussian model 31–41
Generalized homogeneous function 21, 26
Germanium monochromators 84–85
Gibb's free energy 3–4, 10, 20–23, 44
Ginzburg-Landau theory 8–11, 12, 17–19, 22, 29, 35–6, 41, 45, 61, 94, 116–8, 121, 135, 139, 150, 163
Ginzburg-Landau theory of tricritical points 52–4, 135
Goldstone's theorem 6–7

Hamiltonian, Heisenberg 18, 48–51, 105, 108, 140
Hamiltonian, Ising 17–18, 42, 94, 100, 111–120
Hamiltonian, X-Y 18, 29, 61, 94, 100–5, 109, 121, 156, 167

Haldane conjecture 17
Harris criterion 58, 164
Heisenberg antiferromagnet 45–7, 114, 128–134, 145
Heisenberg ferromagnet 44–5, 121–8, 141–3, 155–6
Heisenberg Hamiltonian 18, 48–51, 105, 108, 140
Heisenberg model 18, 29, 44–7, 60, 61, 109, 113, 121–134, 142–3, 149, 164–7
Heisenberg representation 75
Helical structure 152, 156
Helium 6, 18
Helmholtz free energy 8–11, 52–4
Heusler alloy 140, 154–5
History-dependent measurements 169
Holmium 140, 156
Hot source of neutrons 63
Hydrodynamic theory 45–6, 130–2, 149
Hyperscaling 29, 45

Ideal magnet 6
Incoherent cross section 68–9, 170–2
Incoherent scattering of neutrons 68–9, 76
Incommensurate spin ordering 137–9, 149–154, 156, 161
Inequality relationships 14–15
Interaction potential 66, 67
Interrupted phase transition 150–1, 161, 163
Iron 117, 140–9
Irrelevant variables 37, 111
Ising ferromagnet 116–7
Ising antiferromagnets 61, 95–100, 112–6
Ising Hamiltonian 17–18, 42, 94–100, 111–120
Ising model 17–19, 29, 31, 41, 48–51, 58–60, 94–100, 101–120, 160, 164–9
Isothermal susceptibility 11–13, 24
Isotopes, scattering length 68, 170–2
Isotropic system 113, 122, 125
Itinerant electron magnetism 140–156

K_2CoF_4 95–100, 166
$KCuF_3$ 95
K_2CuF_4 95, 100–5, 107
K_2MnF_4 95, 105–8
$KMn_pZn_{1-p}F_3$ 167
K_2NiF_4 95, 105–7
Kosterlitz—Thouless solution 94, 101–5

Lambda anomaly 10
Larmor frequency 91
Laue method 84
Law of corresponding states 16
Length scale 21
Lifshitz point 55–7, 137–9, 154
Lines of critical points 52
Liquid-gas critical phase transition 3–6, 18

Subject Index

LiTbF$_4$ 111, 116–7
Logarithmic divergence 19
Longitudinal fluctuations 106–8, 113–5, 125–7, 146
Long range forces 11, 16, 116–8, 121, 150
Lorentzian form for correlation function 28, 97–9, 109, 120, 124–6, 132, 146–7, 153, 167–8
Lorentzian squared form 168–9
Lower borderline dimensionality 37

Magnetic form factor 71–2
Magnetic scattering cross section 69–72, 76–80, 83
Magnetic scattering of neutrons, 69–72, 76–80
Magnetic structure factor 83
Magnetization 9, 13, 23–4, 70, 94–5, 116–7, 121, 141–3, 163, 157–8
Magnetization density 70, 72
Magnon 44–7, 105, 107, 115, 122, 130–2, 141–5, 151–2, 154–7
Manganese, form factor 72
Maxwellian flux distribution 63, 79
Mean free path of neutron 64
Metastable states 169
Mixed valence 157
Mixing of liquids 6, 18
MnF$_2$ 111, 164–56
MnO 162–3
MnP 137–9, 154–5
Mn$_p$Zn$_{1-p}$F$_2$ 164–9
MnSi 149, 152–4, 156, 161
Model, standard 17–8, 149, 156
Model, solvable 18–9
Mode-mode coupling 45–6
Moderators, neutron 63
Molecular field theory 8
Monochromator 84–5, 90
Mosaic structure 86–8
Multicritical point 52, 56, 59, 135–9

Neutron, energy units 63
Neutron, filter, 85
Neutron, mean free path, 64
Neutron spin echo 90–3, 125, 141, 147
Nickel 140–9
NiO 163
Nuclear forces 63
Nuclear reactors 63
Nuclear scattering of neutrons 67–9, 83
Nuclear structure factor 83

One dimensional systems 19, 60, 94–5, 108–110, 133
Onsager's solution for the two dimensional Ising model 19, 94, 96
Order parameter 6, 80, 118, 132–5, 141, 150

Order parameter, dimensionality 6, 16–19, 132–4, 141
Ordering in alloys 18, 117–120
Ornstein-Zernike correlation function 26–8, 97

Partial differential cross section 65–7, 79, 81
Partition function 31–4, 118
Pd$_2$MnSn 155–6
Percolation 58–60, 166–7
Percolation concentration 58
Percolation critical exponents 59–60, 166–7
Percolation region 58, 164–7
Percolation threshold 58, 167
Phase boundary 3
Polarized neutrons 90–3, 122
Praseodymium, form factor 72
Precession of neutron spin 91
Pseudopotential, Fermi 67
Pyrolitic graphite filter 85

Random fields 60–2, 167–9
Rare-earth metals 140–1, 156
Rb$_2$Co$_p$Mg$_{1-p}$F$_4$ 164–9
Rb$_2$CoF$_4$ 95, 96, 99–100, 105, 164
Rb$_2$CrCl$_4$ 95, 105
RbMnF$_3$ 121, 128–132, 144
Rb$_2$Mn$_p$Mg$_{1-p}$F$_4$ 167
Real space renormalization group 32
Reciprocal lattice 80–3
Reduced temperature 12, 49–50
Relevant variables 37
Renormalization group 31–41, 47, 111, 116, 121, 128, 139, 162
Resolution effects 87–8, 97, 128, 132, 146
RKKY interaction 140, 156
Rotating crystal method 83–4, 128
Rushbrooke inequality 15, 25
Rutile structure 111

Scaling 20–30, 32, 60–1, 113, 131, 133
Scaling, dynamic 42–7, 77, 114–5, 124–6, 129, 146–9, 153–6
Scaling law 22, 120, 143
Scaling theory in the percolation region 60
Scaling theory of crossover effects 49–51
Scattering cross section 66, 83, 170–2
Scattering function 74–8, 95
Scattering geometry 81–3, 85–8, 92
Scattering length 67, 76, 170–2
Scattering vector 67, 72, 95
Schrödinger representation 75
Silicon monochromators 84–85
Single K structure 160–1
Soliton 110
Solvable models 18–19
SO(3) symmetry 18, 132–4
Specific heat 10, 13, 22, 117

Spectral weight function 43, 113, 124, 131–2, 144–7
Spherical model 18, 19, 29
Spin density wave 149–152
Spin diffusion constant 45, 148–9
Spin echo 90–3, 125, 141, 147
Spin flop 55, 136–8
Spin-only scattering 70–3, 83
Spin state of a neutron 66, 70
Spin wave damping 105, 107, 113, 123, 130–2, 143–5, 157
Spin waves 44–7, 105, 107, 113, 115, 122, 130–2, 151–2, 154–7
Spin wave stiffness constant 44–6, 122–4, 141–5, 149
Standard models 17–18, 149, 156, 164
Static approximation 78–80, 95, 97, 128, 141
Stiffness constant of spin waves 44–6, 122–4, 141–5, 149
Strain field energy 118
Structure factor, magnetic 83
Structure factor, nuclear 83
Superconductivity 6, 11
Superfluidity 6, 18
Susceptibility 11, 12–3, 24, 117, 157–8, 160
Symmetry breaking 6, 60–1, 166
Symmetry of order parameter 16–9

Tarko-Fisher correlation function 97–9, 112, 166
Terbium 140, 156
Thermal neutrons 63–80
Three-body interactions 118–120
TMMC 95, 108–9, 133
Total cross section 66
Transition metals 140–156
Transverse fluctuations 106–7, 113–5, 125–7, 145–6, 160
Triangular spin ordering 132–4

Tricritical exponent 53, 135
Tricritical point 51–4, 135–6
Triple-axis spectrometer 90, 141, 147, 150
Triple K structures 159
Triple point 3
Two dimensional systems 94–108, 133–4, 161, 164–7
Type I fcc structure 157, 160–1
Type II fcc structure 161–3

UAs 157–161, 163
UBi 157
UN 157–161
Uniaxial anisotropy 48
Uniaxial term 48
Universality 16–19, 48, 111–2, 116–7, 132–4, 140, 160
UP 157
Upper borderline dimensionality 37, 116
Uranium compounds 140–1, 156–161
US 157–8
USb 157–161
USe 157–8
UTe 157–8

Van der Waal's theory 8, 16
VBr_2 133
VCl_2 133–4

Wavelength of the neutron 63
Weiss molecular-field theory 8

X-Y model 18, 29, 61, 94, 100–5, 109, 121, 156, 167

Z_2xS_1 symmetry 133–4